THE WILEY BICENTENNIAL–KNOWLEDGE FOR GENERATIONS

*E*ach generation has its unique needs and aspirations. When Charles Wiley first opened his small printing shop in lower Manhattan in 1807, it was a generation of boundless potential searching for an identity. And we were there, helping to define a new American literary tradition. Over half a century later, in the midst of the Second Industrial Revolution, it was a generation focused on building the future. Once again, we were there, supplying the critical scientific, technical, and engineering knowledge that helped frame the world. Throughout the 20th Century, and into the new millennium, nations began to reach out beyond their own borders and a new international community was born. Wiley was there, expanding its operations around the world to enable a global exchange of ideas, opinions, and know-how.

For 200 years, Wiley has been an integral part of each generation's journey, enabling the flow of information and understanding necessary to meet their needs and fulfill their aspirations. Today, bold new technologies are changing the way we live and learn. Wiley will be there, providing you the must-have knowledge you need to imagine new worlds, new possibilities, and new opportunities.

Generations come and go, but you can always count on Wiley to provide you the knowledge you need, when and where you need it!

WILLIAM J. PESCE
PRESIDENT AND CHIEF EXECUTIVE OFFICER

PETER BOOTH WILEY
CHAIRMAN OF THE BOARD

Guided Inquiry Experiments for General Chemistry: Practical Problems and Applications

First Edition

Nancy Konigsberg Kerner
University of Michigan at Ann Arbor

Ram S. Lamba
University of Puerto Rico at Cayey

BICENTENNIAL
1807
WILEY
2007
BICENTENNIAL

John Wiley & Sons, Inc.

ACQUISITIONS EDITOR   Stuart Johnson
PRODUCTION EDITOR   Barbara Russiello/Jeanine Furino
MARKETING MANAGER   Amanda Wygal
DESIGNER   Michael St. Martine
COVER PHOTO   Purestock
ILLUSTRATION EDITOR   Sandra Rigby
ILLUSTRATION STUDIO   Dennis Ormond of Fine Line Illustrations, Inc.
MEDIA EDITOR   Thomas Kulesa
WILEY BICENTENNIAL LOGO DESIGNER   Richard Pacifico

This book was set in 12/14 Times by GGS Book Services and printed and bound by Courier / Westford. The cover was printed by Courier / Westford.

This book is printed on acid free paper. ∞

To order books or for customer service please, call 1-800-CALL WILEY (225-5945).

Library of Congress Cataloging-in-Publication Data:

Kerner, Nancy Konigsberg.
  Guided inquiry experiments in general chemistry : practical problems & applications / by Nancy Kerner and Ram Lamba—1st ed.
    p. cm.
  ISBN 978-0-471-69842-5 (pbk.)
  1. Chemistry—Laboratory manuals. I. Lamba, Ram. II. Title.
  QD45.K385 2008
  542—dc22

                              2007031823

ISBN-13   978-0-471-69842-5
ISBN-10   0-471-69842-3

Printed in the United States of America

10 9 8 7 6 5 4 3 2 1

Preface

GOAL

Our goal for writing this laboratory manual was to provide an exciting and meaningful laboratory experience for students by challenging them to solve a practical problem ("Why Did My Watch Stop Suddenly?" "How Deep Can a Diver Go?") by drawing conclusions from experimental data. Each experiment ends with questions requiring feedback on the solution to the problem in the form of a report. A closing "Extension and Applications" section emphasizes the practical applications of chemistry by asking questions that require students to apply and extend experimental results to untested systems and real-world situations.

Guided Inquiry Experiments for General Chemistry: Practical Problems and Applications has been developed from our belief that it is valuable to use the laboratory as a process experience where students develop understanding of key chemical concepts from experiment. The approach encourages scientific thinking in which the emphasis is on asking "what do the data mean?" rather than on making "chemically correct" deductions. This approach also more closely reflects what occurs in scientific research. After conducting the labs, students report improvement in their communication and inquiry skills, including forming and testing hypotheses, analyzing data, and designing experiments.

INTENDED USERS, LEVEL, AND BREADTH

This manual covers two semesters (or three quarters) of a general chemistry program where a student can expect to spend three hours per experiment in the laboratory; extensive analysis and/or discussion of the implications of the data may lengthen this time.

The experiments may accompany any general chemistry text and are sorted into units based on topics common to most texts. Each unit offers several experiments on a given topic, where each focuses on a different key chemical concept. It is possible for a user to perform selected individual experiments in any order that best fits the background of students.

EXPERIMENT DESIGN

The experiment design is based on accepted educational theory that students construct their knowledge from involvement, experience, and models. The general framework of the experiments is inquiry (introduce a problem, collect data to solve the problem, organize and analyze data, apply and/or evaluate data). To further involve students in the inquiry process, guiding questions focus on the interpretation or "what if" of nearly every part of the laboratory procedure. We provide information, models, and guidelines rather than detailed procedures or explicit "recipes" to encourage active student involvement. Student dialogue and collaboration are encouraged in order to maximize the contributions that peer collaboration and teamwork can make in improving understanding. Students are often directed to compare and share observations and data to gather a large set of data that the students then analyze.

SAFETY GUIDELINES AND TECHNIQUES

Students are not expected to know laboratory safety guidelines or "construct" skills or techniques. Experiments in the introductory unit (Scientific Method) provide explicit directives and models for basic skills, including designing tables for data collection and guidelines for drawing graphs. Details are streamlined in subsequent experiments to promote active learning. Caution statements are placed in the text of the experiment procedure at appropriate positions to ensure that students are aware of immediate safety procedures.

EXPERIMENT CONTENTS

Introduction. One or more paragraphs establish the problem and experiment goals. An opening photograph reinforces the focus of the experiment.

Objectives. Students' goals are detailed with regard to solving the problem, performing the experiment, and collecting and analyzing data.

Procedures. A short paragraph provides an overview of the lab procedures.

Guidelines. We include guidelines for performing the experiment and collecting the data rather than explicit procedures to more actively involve the student.

Team Data. Directives or tables are provided for recording and organizing individual team data.

Compiled Team Data. In this section students organize the data from different teams.

Optional Extension. This optional problem challenges students to predict and experimentally test the impact of varying testing conditions on results.

Data Analysis and Implications. These are guided directives for interpreting data and ''what if'' questions regarding the data implications with regard to the problem to be solved.

Experiment Report. These directives for the laboratory report typically request the student to provide a summary of conclusions and recommendations with regard to the investigated problem and also include explicit reference to data collected and analyzed during the experiment.

Questions: Extensions and Applications. These are questions that require students to apply and / or extend results to untested samples and new situations.

TO THE STUDENT

Chemistry is an experimentally based science. We know what we know because many scientists have made millions of experimental observations over several hundred years. From these observations, fundamental understandings and principles have been realized regarding the properties and reactivity of matter. In turn, chemical understanding and principles can be used to solve real-world problems dealing with the properties and reactivity of matter.

The guided inquiry experiments in this manual will involve you in the process of looking for property or reactivity data patterns to solve problems. The experiments will also acquaint you with some of the processes (forming hypotheses, designing experiments, collecting and interpreting data, and explaining results) used by chemists to investigate chemical phenomena.

No one is born knowing how to perform chemistry labs (not even the instructors). Unlike some traditional laboratory approaches, you do *not* need to know anything about the problem being investigated in order to do well. You *do* have to prepare in advance for the labs, work conscientiously and safely during the lab period, and think about the experiments that you are performing.

When conducting research to solve a problem, chemists do not know the outcome in advance of performing the experiments. In contrast to some traditional laboratory experiments, here you are not expected to know the answer to the problem being investigated. If the outcome is known in advance there is no problem to solve. As with any research, there are a number of interpretations that will fit a data set. The ''right answer'' is any response that follows logically from the data and that clearly states your reasoning. An answer solely from a reference or textbook is unacceptable. Using information from a reference or textbook to support your data-based answer is acceptable.

Inquiry investigations consist of three phases: exploration, organization, and application. During exploration you will gather information (data) in order to answer the problem under investigation. During the organization phase, you will manipulate the data to look for patterns. Science consists, in large part, of attempts to arrange apparently diverse data so as to discover an underlying systematic order. The periodic table is an example of such a systematic order, where the elements are arranged by increasing atomic number and display a relationship between atomic number, structure, and properties. In the laboratory, you might consider whether the color of a solution can be linked to the structure of the species in the solution. If there is a link between color and structure, you should be able to make a rational prediction about the color of a sample you've never seen.

You can expect to experience confusion as you perform the labs and seek answers to problems. Confusion is a signal that you need to invest time sorting through the information and thinking about the implications of the results. Research indicates that your involvement in the process of resolving the confusion will facilitate development of your reasoning skills.

You will do most of your experiments as a member of a team. You and your team will combine and compare data, instead of competing for the ''right'' answer. Research has shown that students learn better, develop interpersonal skills, and enjoy a course more in a group (rather than individual) learning environment. In addition, teamwork typifies real-world science better than independent learning. Team learning does not mean that students simply work side by side on a problem or the best student works while the others watch. Rather, a well-functioning group is interdependent.

INSTRUCTOR'S MANUAL

A *Teacher's Manual* (available to instructors from Wiley) provides detailed information for each experiment and includes:

- An overview of the experiment
- Key chemical concepts the experiment develops
- Techniques/skills needed to perform the experiment
- Equipment and reagents needed per 24 students
- Reagent preparation including indicators, indicator substitutes, and unknowns
- Sample data
- Pre- and post-lab implementation suggestions
- Possible answers to questions
- Details for alternative low-cost equipment when applicable

To access the *Teacher's Manual*, please visit www.wiley.com/college/kerner. Click on the Instructor Companion Site to log-in and obtain your instructor password to view and/or download the manual.

LABORATORY EQUIPMENT

Only simple laboratory glassware and equipment are necessary for completing most experiments. In addition to balances, spectrophotometers (Experiments 3-1 and 3-3), pH meters (Experiments 9-1 and 9-2), and pH electrodes and voltmeters (Experiments 11-1 and 11-2) are suggested. Low-cost alternatives to commercial pH electrodes using carbon (graphite) electrodes are described in the *Teacher's Manual*. If the instrumentation is unavailable, these experiments can be modified or omitted.

ACKNOWLEDGMENTS

This manual is the result of several projects funded primarily by the National Science Foundation and developed chiefly at the University of Puerto Rico at Cayey, Inter American University of Puerto Rico, Metropolitan Campus, and the University of Michigan at Ann Arbor. The experiments have been thoroughly tested with undergraduate premed, engineering, and science (including chemistry), as well as non-chemistry, non-science, and future teachers majors. Successful use of the experiments has occurred outside the institutions of origin, including secondary institutions, community colleges, and universities.

We would also like to acknowledge the contributions of George Bodner from Purdue University, Jim Spencer from Franklin and Marshall College, and Cathy Middlecamp from University of Madison at Wisconsin to preliminary drafts of many of these experiments. We cannot thank Jim Spencer enough for his ceaseless encouragement to move forward with this project and help promote acceptance of the guided inquiry model and the methods of the project. We would also like to acknowledge the assistance of Carl Berger and graduate students in chemistry and education from University of Michigan at Ann Arbor, in the evaluation of the impact of preliminary drafts of some experiments on student attitudes and cognitive skills.

We also wish to recognize the contributions of Jairo Pardo and Nelson Granda, both from University of Puerto Rico at Cayey, and Agnes Dubey from the Metropolitan Campus of Inter American University of Puerto Rico for testing many experiments, giving feedback, and in some cases for the development of data for the *Teacher's Manual*. Special thanks are due to Shiva Sharma, a former graduate assistant at Inter American University of Puerto Rico, who tested and retested many of the original drafts of experiments with high school and undergraduate students. Special thanks are also due Ramon de la Cuetara from Inter American University of Puerto Rico for his collaboration and input on the construction and use of the low-cost equipment. We are indebted to the many undergraduate students who continuously gave us feedback toward improving the experiments, and, at the same time, giving us assurance that the approach helped them think more critically.

It is a pleasure to acknowledge the support, encouragement, and assistance of the staff at John Wiley and Sons for this guided inquiry lab manual. Thanks to Acquisitions Editor Kevin Molloy and Editor Stuart Johnson, for their commitment to this manual. Thanks to Hilary Newman, Photo Editor at Wiley, for assistance in obtaining the photographs for this edition; Barbara Russiello, Production Editor, for her attention to the design of the manual; Jeanine Furino, Production Editor, for her attention to managing the production, and to Cathy Donovan for attention to countless details.

Suggestions for revisions of preliminary drafts for these experiments were guided by student and instructor feedback and contributions from members of NSF Advisory Committees to the University of Michigan at Ann Arbor, University of Puerto Rico at Cayey, and Inter American University of Puerto Rico. Advisory Committee members included Rick Moog and Jim Spencer, Franklin and Marshall College; Mauri Ditzler, Millikan University; Joel Russell, Oakland University; and Michael Abraham; The University of Oklahoma.

We greatly appreciate feedback from the following reviewers:

Kristine Kirk
College of Notre Dame of Maryland

Kathie Snyder
Winthrop University

Roderick M. Macrae
Marian College

Margaret Kerr
Worcester State College

Thomas Burkholder
Central Connecticut State University

Kathleen Brunke
Christopher Newport University

Donnie Byers
Johnson County Community College

Patrick Barber
Longwood University

Edward Baum
Grand Valley State University

Cathy Middlecamp
University of Wisconsin at Madison

Kristine Kirk, College of Notre Dame of Maryland, served as lead reviewer, and her detailed feedback and guidance are especially valued.

REVIEWER COMMENTS

"There are few inquiry-based lab manuals available, and this one is perhaps the best I've seen. The use of scenarios to set each experiment in a context is engaging. The format appeals to students and helps them develop knowledge and skills that many instructors claim are essential, but few students actually learn and few instructors actually teach in traditional lab courses. The emphasis on science writing skills is a plus. Another plus is the emphasis on teamwork in the experiments, and collaborative work to gather a large set of data that the students then analyze."
Kristine Kirk, College of Notre Dame of Maryland

"I like the added real-world framework. The Extensions/Application sections are excellent, and would I think have a very positive impact on learning."
Roderick M. Macrae, Marian College

"This is a great way to teach lab to general chemistry students. It emphasizes a more realistic way in which a scientific problem is approached, and requires students to think, interpret and draw conclusions from data they obtained."
Kathie Snyder, Winthrop University

"I believe that students would be the real winners in a laboratory approach like this. They get to see the relationships and develop an understanding of the content rather than just memorize information. The approach is truly making the lab an extension of the lecture. They are not just validation labs which in my opinion are the least effective means of teaching."
Donnie Byers, Johnson County Community College

"This type of inquiry-based chemical laboratory experimentation will improve student's thinking and problem solving skills. It will also give them a sense of accomplishment when they complete a lab and its report especially as they must pretend that they are part of a company team and must make an oral / written report with supported recommendations to the group."
Patrick Barber, Longwood University

"This project is well thought-out, well-organized, and high quality. It is clear that the authors took the time to carefully design and test these labs. The risk you run with this material is that users who are not familiar with guided-inquiry instruction might be tempted to "improve" the labs by making ill-considered changes and additions. With a properly documented lab manual, the authors can take advantage of a valuable opportunity to educate their peers and help promote increased acceptance of the method."
Edward Baum, Grand Valley State University

We invite suggestions from colleagues, instructors, and students.

Nancy Konigsberg Kerner
University of Michigan at Ann Arbor

Ram S. Lamba
University of Puerto Rico at Cayey

Contents

UNIT 1 SCIENTIFIC METHOD

Experiment 1-1 How Long Can a Bubble Last? 1

Experiment 1-2 Are All Pennies the Same? 19

UNIT 2 STOICHIOMETRY

Experiment 2-1 How Much Hydrogen? 39

Experiment 2-2 How Much Is Too Much? 55

UNIT 3 SOLUTIONS, ATOMIC STRUCTURE, AND PERIODICITY

Experiment 3-1 What Makes a Solution Colored or Colorless? 67

Experiment 3-2 Are There Property Patterns? 87

Experiment 3-3 What Factors Affect Color Intensity? 107

UNIT 4 REACTIVITY AND PERIODICITY

Experiment 4-1 What Factors Affect the Solubility of Ions? 121

Experiment 4-2 Can Toxic Ions Be Removed from Water by Precipitation? 143

UNIT 5 GASES

Experiment 5-1 How Do Gas Volumes Change as We Ascend and Descend? 163

UNIT 6 THERMOCHEMISTRY

Experiment 6-1 Which Salts Make Good Cold Packs and Hot Packs? 183

Experiment 6-2 How Is Heat Measured Indirectly? 201

UNIT 7 STRUCTURE AND PROPERTY RELATIONSIPS

Experiment 7-1 Do Like Repel or Attract? 225

UNIT 8 EQUILIBRIUM

Experiment 8-1 How Can a Chemical Equilibrium Be Shifted? 241

Experiment 8-2 How Do Soaps Compare in Controlling Malodor? 261

UNIT 9 ACIDS AND BASES

Experiment 9-1 Is It Acidic, Basic, or Neutral? 281

Experiment 9-2 Are Acid-Base Properties in Consumer Products Predictable? 299

UNIT 10 OXIDATION AND REDUCTION

Experiment 10-1 Do Chemicals Duplicate the Effects of a Battery? 317

Experiment 10-2 Which Metal Is Appropriate for Food and Beverage Storage? 337

UNIT 11 ELECTROCHEMISTRY

Experiment 11-1 Which Metals Provide the Best Voltage? 353

Experiment 11-2 Why Did My Watch Stop Suddenly? 367

Experiment 11-3 Why Is a "Harley" Chrome-Plated? 387

UNIT 12 KINETICS

Experiment 12-1 How Long Will It Take? 401

Experiment 1–1

How Long Can a Bubble Last?

You have recently been placed on a research team to help provide information for your company regarding their soap product. One customer recently returned the soap solution, claiming that the liquid produces bubbles that burst instantaneously. Your team's task is to carry out experimental procedures that can characterize the conditions needed to produce lasting bubbles.

- Investigate the effect of soap concentration on the time that a bubble lasts.
- Design an experiment on bubble size and bubble longevity on different surfaces if time allows.
- Create a graphical plot of your bubble data to visualize and interpret the results.
- Report your findings and recommendations to the company.

For purposes of comparison and validity of results, you need to be able to blow a single bubble of a given size from your soap solution. The information listed below indicates how to prepare your soap solution. Before starting your investigations, practice producing bubbles of consistent size as indicated in Preliminary Testing below.

INFORMATION

- To prepare a dilute soap solution, add 4 mL of Dawn™ liquid soap to a 100-mL volumetric flask and then dilute to the mark with water.
- If a magnetic stirrer is available, add a stir bar and slowly stir the solution for two to three minutes. You must stir slowly to minimize the formation of bubbles.
- Transfer the prepared soap solution to a beaker or Erlenmeyer flask.

PRELIMINARY TESTING

The surface on which the bubble is blown must be flat, clean, level, and away from drafts. Draw a circle on the surface that is 5 cm in diameter. You might achieve this by cutting out a 5-cm circle from a paper plate and then tracing the circle edge with a pencil. Obtain or prepare a dilute soap solution as indicated in the information above. Add 1 mL (20 drops) of the dilute soap solution to the center of your drawn circle. Place the tip of a hollow plastic coffee stirrer or a small straw beneath the surface of the soap solution. Blow *gently* and *continuously* into the open end of the coffee stirrer or straw to form a single bubble. (If more than one bubble forms when you blow through the straw, carefully clean the surface with a damp cloth so that no soap solution remains and start over.)

> *Caution:* Do not allow the soap solution to enter your mouth!

Use a stop or digital watch with increments of seconds to record the instant that the bubble becomes large enough to fill the circle. Then record the instant at which this 5-cm-diameter bubble bursts. One team member can blow the bubble and another record the lifetime of the bubble. Take turns and practice this measurement several times, until you have obtained and recorded consistent results.

Time (s) Bubble = 5-cm Circle	Time (s) Bubble Bursts	Lifetime (s) of 5-cm Bubble

PART 1— THE EFFECT OF CONCENTRATION

What will happen to the lifetime of a bubble as you increase the concentration of the soap solution? You are to study four different soap concentrations (provided unless indicated otherwise) with a minimum of 4 mL to a maximum of 10 mL of Dawn™ diluted with water to a total volume of 100 mL.

1. Scientists generally start an investigation with a hypothesis. The hypothesis needn't be correct and can be disproved by experiment. What do you expect to happen to the lifetime of the bubble as you increase the concentration of the soap solution? Record your hypothesis.

Hypothesis

2. Read the following experiment guidelines and then proceed to measure the time it takes for a bubble of a given size (5 cm) to burst using different soap concentrations.

GUIDELINES

- For the purpose of sharing results, teams must use the same size bubbles, type of surface, and soap solution temperature for bubble production.
- You may share and compare the measurements taken at room temperature in Table 1 with those of other teams provided you use the same type of surface and size of bubble.

TEAM DATA

Bubble size (cm): _____ Surface type: _____ Temperature (°C): _____

Table 1 The Time a Bubble Lasts versus Soap Solution Concentration

Trial	_____ mL Soap/100-mL Solution	Time (s)
1	4	
2	6	
3	8	
4	10	

Optional Extensions

1. If the soap solutions were prepared using varying water sources (i.e., distilled vs. hard vs. soft water) would your results differ?
2. Does the lifetime of a bubble depend on soap solution temperature? For example, if the soap solution temperature were higher, would your results differ?

Data Analysis and Implications (Part I)

1. Create a graphical plot of your data regarding the lifetime of a bubble versus soap concentration.

   Note: The custom in drawing graphs is to use positions on the horizontal axis (*x* axis) to represent values of the independent variable, and to use positions on the vertical axis (*y* axis) to represent the dependent variable (calculated or measured values). Thus for this experiment the *x* axis is soap concentration (_mL soap/100 mL) and the *y* axis is lifetime of the bubble (seconds). Make sure your graph has properly labeled axes, is titled, and identifies the members of your team. Attach your graph to this report.

2. Post your results on the board (or as indicated by your supervisor). Compare your results with those of other teams. Do the graphs relate to one another in any systematic way? Describe any consistent or inconsistent pattern you observe.

3. Based on the posted graphs, what actually happened to the lifetime of the bubble as the soap solution became more concentrated?

4. Did the results support your predictions? Why or why not?

Does the size of a soap bubble affect its lifetime? Does it matter whether a customer uses a particular type of surface to produce bubbles? Your instructor will ask different teams to investigate one of these two questions and share the results. Part IIA—The Effect of Size, and Part IIB—The Effect of Surface, ask you to design your experiment and collect data to provide an answer to these questions. You may include data already recorded in Part I for comparison purposes if your testing conditions such as soap concentration are identical.

PART II— EXPERIMENT DESIGN: THE EFFECT OF SIZE AND SURFACE

Would a change in the size of the bubble affect its lifetime? If so, how? Your company requires you to report results for four varying sizes of bubbles with diameters between 3 and 7 cm.

Part IIA— The Effect of Size

1. What do you expect to happen to the lifetime of a bubble as you increase the diameter of the bubble? Record your hypothesis.

   Hypothesis

2. Read the experiment guidelines below and then proceed to design and conduct your experiment. Record your results for the effect of size on bubble lifetime in Table 2.

GUIDELINES

- Team trials must be conducted using controlled conditions (constant soap concentration, soap solution temperature, and surface type).
- Measurements already taken in Part I may only be used for comparison purposes if the same testing conditions are used.
- Teams collaborating on data collection must use the same testing conditions.
- Repeat the experiment procedure until you get consistent results.

PROCEDURE AND DATA

Table 2 The Time a Bubble Lasts versus Size of the Bubble

Trial	Diameter of Bubble (cm)	Time (s)
1		
2		
3		
4		

Soap concentration (mL/100 mL): _____

Soap temperature (°C): _____   Surface type: _____

3. Record and compare the measurements from different teams using the same testing conditions.

Optional Extension   Some companies add additives such as glycerin to make commercial bubble products. Design experiments to test whether addition of glycerin to the soap solution alters your results.

Data Analysis and Implications (Part IIA)

5. Prepare a graph of your data regarding bubble lifetime versus the initial size of the bubble and attach it to this report. What do the results imply?

6. Provide a record (preferably a table) of the compiled data from different teams taken under the same testing conditions regarding bubble lifetime versus the initial size of the bubble. Be sure to specify the conditions of testing.

7. Based on the combined data, what actually happened to the lifetime of the bubble as the size of the bubble increased? Do the results support your predictions? Why or why not?

Does it matter whether a customer uses a particular surface (e.g., wooden, enamel, glass, plexiglass table top, or other surface) to produce bubbles? You are to study the effect of at least one different surface type on the time it takes a bubble to collapse.

1. Record your prediction (hypothesis) regarding the time it takes for a bubble to collapse on the particular surface.

Hypothesis

2. Read the guidelines provided below, design your experiment procedure, and record your data in Table 3.

GUIDELINES

- Team trials must be conducted using controlled conditions (constant soap concentration, soap solution temperature, and bubble size).
- Measurements already taken in Part I may only be used for comparison purposes if the same testing conditions are used.
- Teams collaborating on data collection must use the same testing conditions.
- Repeat the experiment procedure until you get consistent results.
- A glass or plastic crystallizing or Petri dish may be used in place of a flat sheet, plastic, or glass surface.

PROCEDURE AND DATA

Table 3 The Time a Bubble Lasts versus Surface Type

Trial	Diameter of Bubble (cm)	Time (s)
1		
2		
3		
4		

Soap concentration (mL/100 mL): _____

Soap temperature (°C): _____   Surface type: _____

Bubble size (cm): _____

3. Record the measurements from different teams using different surfaces and the same testing conditions.

Optional Extensions

1. Design experiments to test the effect of surface curvature on bubble production. For example, if a surface is depressed (e.g., the bottom of an inverted beaker) rather than flat, will bubble production alter?

2. Design experiments to determine whether the temperature of the soap solution affects the results on your surface type.

Data Analysis and Implications (Part IIB)

8. Provide a record (preferably a table) of data collected from different teams using different surfaces to produce bubbles. The compiled data must be for production of bubbles of the same size (cm) where the soap is of the same concentration and temperature. Your record should indicate the specific testing conditions.

9. Based on the combined data of the different teams, does the type of surface influence the rate of bubble breaking? How?

10. Do the results of the other teams support your predictions? Why or why not?

Experiment 1–1   Report

Write a report to your company summarizing your findings with regard to the problem under investigation—How long can a bubble last? Suggest a response to the customer who returned the soap solution claiming that the "liquid produces bubbles that burst instantaneously." If you feel the answer to the problem is inconclusive or unclear, suggest other investigations the company might wish to pursue. Refer to your observations, data, and graphs to support your conclusions.

Questions 1–4 are based on the team data shown in Table 4, which resulted from a controlled experiment investigating the effect of temperature on the lifetime of a bubble.

Table 4 The Time (s) a Bubble Lasts versus Temperature (°C)

Temperature (°C)	11	17	23
Average time (s)	122	131	139

1. What experimental variables did the team control when designing the experiment to investigate the effect of temperature on the bubble lifetime? For example, would the results be valid if the team failed to use a constant soap concentration?

2. Which of the following hypotheses do the temperature study results support?

   a. The lifetime of a bubble decreases with increasing temperature.
   b. There will be a direct (straight-line) relationship between the lifetime of the bubble and the temperature.
   c. The lifetime of the bubble is dependent on concentration.

3. Based on the above temperature data and your concentration data, what are the optimal conditions needed to produce lasting bubbles? For example, would you recommend that customers use concentrated or dilute soap solutions and higher or lower than room temperature soap solutions to produce lasting bubbles?

4. It is a scientific fact that the vapor pressure of water inside a bubble increases as the temperature of the solution increases (and keeps it moist). Are the temperature study results consistent with the hypothesis that differences between the vapor pressure inside and outside the bubble play an important role in determining the lifetime of the bubble?

5. Companies often add additives to soap solutions for commercial bubble products. Search the library or the Web and determine what additives are used in making commercial bubble products. Indicate how to design experiments to test these additives to determine whether the bubble product can be improved. Are the additives present to improve bubble production or for some other reason such as product appearance?

Are All Pennies the Same?

INTRODUCTION

Customers deposit their excess pennies in a "penny cup" at the fast-food chain where you work. You have been collecting these pennies over a period of years. Your uncle has indicated he will pay you fifteen dollars for every ten dollars in pennies you give him provided he can select the pennies. Your uncle's offer arouses your curiosity and causes you to ponder whether all pennies are the same, regardless of age. Your task is to determine if pennies of different ages are alike or different in properties.

OBJECTIVES

- Become familiar with experimental skills and methods that may be used to determine whether properties of different samples (e.g., pennies of different ages) are alike or different.
- Convert the tabulated property data into a graph to visualize and interpret the results.
- Determine whether the properties of pennies are the same, regardless of age.
- Determine whether the ages or number of investigated pennies has an impact on the findings.

PROCEDURES

You and your peers will work in teams so as to collect property data rapidly on multiple pennies for each year over a number of years. You may wish to bring a jar of your collected pennies to lab.

**PART I—
QUALITATIVE DATA**

Before conducting research and collecting quantitative data, scientists generally record a hypothesis and collect qualitative observations of research samples. You will conduct qualitative observations of properties (e.g., color, wear, relative thickness, and shape) of pennies of different ages.

1. Will your observations indicate that all pennies are the same, regardless of age, or will your observations point to differences

among the pennies, based on age? Record your hypothesis with regard to this research question. The hypothesis need not be correct and can be disproved by the data you will collect. For example, you might imagine (hypothesize) that pennies of different years will not be distinguishable or will be alike with regard to one or more qualitative characteristics such as color and/or thickness.

Hypothesis

2. Obtain 3–6 pennies of different years (from the present back to the 1960s or earlier). Record your qualitative observations of the pennies in Table 1.

Table 1 Qualitative Properties of Different Pennies

Sample #	Year of Mint	Qualitative Observations

Data Analysis and Implications (Part I)

1. Do your recorded observations of properties (Table 1) show significant variation such as to indicate differences among the pennies based on age? For example, does the color or relative mass or shape of a penny appear to change at specific ages?

2. Did your observations support your hypothesis? Why or why not?

1. Is the mass of all pennies the same, regardless of the age (year of mint) of the penny? Before starting your study, record a hypothesis with regard to this question. If your hypothesis is supported by your data, what will you quantitatively observe as you measure the mass of pennies of increasing age?

Part IIA—
Mass Measurements

Hypothesis and anticipated observations

2. Measure the mass of different pennies following the guidelines provided below.

GUIDELINES

- Each team should determine and record the mass to $+/-0.01$ gram for 3 or 4 pennies.
- Record your team's mass data in the form of a table (Table 2). Record the column headings indicating the year of mint and mass (g) of the pennies.

TEAM DATA

Table 2 Year of Mint versus Mass (g) of Pennies

3. Create a graph (see the following information) of your team data. Post your results as indicated by your instructor for sharing different team results.

INFORMATION

- The horizontal (*x*) axis should depict values of the independent (year of mint) variable.
- The vertical (*y*) axis should represent the dependent (mass of pennies (g)) variable that may be calculated or measured values.
- The graph should be titled, have labeled axes, and indicate the names of the experimenter/s (research team member/s).

4. Collect the mass data from the different teams using a computer or Table 3 provided below. Make sure to record column headings in Table 3.

5. Prepare a graph using the compiled team data. Post your graph as indicated by your instructor.

COMPILED TEAM DATA

Table 3 Year of Mint versus Mass (g) of Pennies

Data Analysis and Implications (Part IIA)

3. Compare the graph of your individual team results with the graph of the compiled team data and answer the following questions:

   a. Does the sample size affect the findings? Specifically, compare the results from the graph based solely on your team data to the results based on multiple team data.

   b. Do the compiled results of the different teams support or contradict your hypothesis? Why or why not?

   c. Based on the compiled team data, can you predict the average mass of a penny if you know its year of mint? Why or why not?

Materials that are alike have the same density. Is the density of different pennies the same regardless of the age of the penny? Density is defined as the ratio of an object's mass to its volume and is typically expressed in g/mL or g/cm^3. To determine an object's density we need to know both its mass and volume. In order to make some observations about the volume of pennies of varying age, we will select pennies of similar mass (+/−0.01g) and measure their volume. The volume of pennies will be determined by water displacement. It will be necessary to use a minimum group of three pennies in order to appropriately measure the volume of water displaced. Your instructor will assign your team to a group of 3, 6, 9, or 12 pennies.

Part II B—
Volume Measurements and Density Determination

1. Is the density of different pennies the same, regardless of the age of the penny? Record your hypothesis with regard to this question.

   Hypothesis

2. Choose a group (3, 6, 9, or 12 assigned by your instructor) of pennies of similar mass (+/− 0.01 g) that you have weighed to +/− 0.01 g. Record your data below and the method used to select your pennies of similar mass.

   Selection of Group of Pennies   Group size: _____

3. Measure the total mass of the team-selected group of pennies. Record the oldest to newest mint dates in Table 4 and calculate the average mass per penny.

TEAM DATA

Table 4 Mint Years and Average Penny Mass

Year of Mint (oldest–newest)	Number of Pennies	Total Mass (g)	Average Mass (g)

4. Measure the total volume (mL) of your team-selected group of pennies by measuring the amount of water they all displace when added to a 50-mL graduated cylinder containing a known volume (about 40 mL) of water.

5. Reproduce the data collected thus far (total mass, average mass, total volume) for your team-selected group of pennies Table 5. Calculate the average volume per penny and the average density (g/mL).

6. Collect and record the data of all the different teams in Table 6.

Table 5 Penny Mint Years versus Average Volume and Density

Year of Mint (oldest–newest)	Number of Pennies	Total			Average		
		Mass (g)	Volume (mL)	Density (g/mL)	Mass (g)	Volume (mL)	Density (g/mL)

COMPILED TEAM DATA

Table 6 Penny Mint Years versus Average Mass, Volume, and Density

Year of Mint (oldest–newest)	Number of Pennies	Total Mass (g)	Total Volume (mL)	Average Density (g/mL)
	3			
	3			
	3			
	6			
	6			
	6			
	9			
	9			
	9			
	12			
	12			
	12			

Data Analysis and Implications (Part IIB)

4. Is penny volume independent of or dependent on the year of mint of a penny? Refer to the volume data in Table 6 to support your answer.

5. Create a graphical plot of the total penny mass versus the total penny volume using the compiled data from different teams in Table 6 and attach your graph to this report. Use the results to answer the following questions.

   a. Are the results consistent with your hypothesis? Why or why not?

   b. Does sample size affect findings? Specifically, compare the results based solely on your team data for the investigated properties to the findings based on multiple team data.

c. Calculate the slope ($\Delta y/\Delta x$ in g/mL) of the line in the compiled graph. Compare the resulting value with the calculated data in Table 6.

d. Can you predict the density of a penny if you know its year of mint? Refer to the data to support your answer.

Is the chemical reactivity of pennies the same, regardless of the age of the penny?

1. Record a hypothesis regarding the reactivity of pennies of different age.

   Hypothesis

2. Use a metal file to remove some of the metal at five to six different points on the outer edge of your oldest and newest penny. (If you aren't sure whether enough metal has been removed, ask your instructor.) Does a given penny appear to be heterogeneous or homogeneous in composition based solely on your visual observations? Record your observations.

3. Obtain a pair of beakers. Label each beaker with the year of penny mint of your oldest and newest penny. Carefully pour 40–50 mL of 3 – 4 M hydrochloric acid into each beaker.

   > *Caution:* Hydrochloric acid causes burns! Avoid acid splash! Don't drop the penny into the beaker! If acid comes in contact with your skin or clothing, wash the area profusely with water!

4. Use a tweezer to grasp the pennies and **cautiously** slide one of the pennies into each acid solution as shown in Figure 1. *Don't drop the penny into the beaker because you might splash your skin or clothing.* Record your observations.

   Observations

Figure 1 Inserting a Penny Into Acid

5. Observe and compare your reactivity results (beaker contents) with those from other teams.

6. When you have completed your observations, dispose of the beaker contents following the waste disposal guidelines provided below.

Waste Disposal Guidelines

> *Caution:* Do not dump the 3 – 4 M acid down the drain! Dilute the acid with 10 times the volume of water and drain-dispose! Remove the penny with tongs or forceps! Do not remove the penny with your fingers!

When you have completed your observations, obtain a large beaker of water with about 400 mL of water. Slowly pour the acid into the water. Remove the penny with tongs or forceps and rinse. The diluted acid is safe to pour down the drain. Dispose of the used penny as directed by your instructor.

Optional Extension   Cover each beaker with a watch glass and set them aside (as indicated by your instructor) to observe during the next laboratory period.

Data Analysis and Implications (Part IIB)

6. Do your observations support your hypothesis regarding reactivity and age of a penny? Why or why not?

7. Properties such as the density of metals are listed in reference books such as the *CRC Handbook of Chemistry and Physics*. The densities of some common metals that appear silvery in appearance are: aluminum = 2.70 g/cm^3; silver = 10.5 g/cm^3; zinc = 7.11 g/cm^3; nickel = 8.90 g/cm^3; whereas the density of copper = 8.92 g/cm^3. Based on this information and your observations, does the interior of some copper pennies contain some metal other than copper? If so, what is the most likely identity of the metal? Refer to the data collected in this investigation to support your answer.

Experiment 1–2   Report

Summarize your findings with regard to the question: Are all pennies the same regardless of age? Include a recommendation regarding whether you should accept your uncle's offer to pay fifteen dollars for every ten dollars in pennies you give him providing he can select the pennies. Refer to specific class data in the tables or graphs to support your conclusions and recommendations.

1. The density of copper is 8.92 g/cm^3. Does a newer penny have a higher, lower, or the same percentage of copper? How do you know? Refer to the compiled team data.

2. Nickels that are pre-1959 are worth more money. In particular, nickels from 1942 through 1944 are especially valuable. Search on the Web to determine why these particular ages of nickels are precious. Are the properties of nickels from 1942 through 1944 different from those of a current nickel and are such differences measurable? If so, how?

3. Table 7 shows data collected regarding human body density versus percent of body fat. Human body density is determined from body mass and volume. The volume of a human body is determined by measuring the amount of water displaced when a person is submerged in water. The percentage of body fat is determined by how buoyant the body is in water. The density of human fat is 0.093 g/mL, whereas that of water is 1.00 g/mL.

Table 7 Human Body Density versus Percent of Body Fat

Body Density (g/cm^3)	% of Body Fat
1.02	33.9
1.04	25.25
1.06	16.95

   a. Will a person with a higher percentage of body fat be more or less buoyant in water?
   b. Do the results support a hypothesis stating: Body density is independent of percentage of body fat? Why or why not?
   c. Estimate the percentage of body fat for a person with a body density of 1.034 g/cm^3.

4. Cans of Diet Coke^{TM} float and cans of Coke^{TM} are identical in shape and volume, and yet cans of Coke^{TM} sink in water while cans of Diet Coke^{TM} float (density water = 1.0 g/mL). Compare the mass of artificial sweetener and mass of sugar in these products.

5. A 55-gallon drum holds 309.8 pounds of gasoline or 459.0 pounds of water. (a) Which is more dense: gasoline or water? (b) Gasoline and water do not mix; rather, they form two separate phases. Which liquid will be on top and which will be on the bottom?

6. The iron in steel has a density of 7.86 g/cm^3. When iron rusts, it forms hydrated Fe_2O_3 that has a density of 5.12 g/cm^3. Which occupies a larger volume: a gram of iron or a gram of rust? If a sample of iron rusts, will the volume occupied by the rust be larger or smaller than the initial volume of the reacting iron?

Experiment 2–1

How Much Hydrogen?

INTRODUCTION

A local company has recently acquired a large supply of magnesium and aluminum and has asked you to explore gas production using these metals. The company uses hydrogen gas in various processes and does not have personnel familiar with the chemistry of gas production, which currently uses zinc metal and acid. The firm wants you along with appointed teams of chemists to measure and compare the hydrogen gas generated by the different metals. They want to know whether magnesium or aluminum is a reasonable alternative to zinc for producing a given volume of hydrogen gas.

OBJECTIVES

- Successfully measure and compare the amount of hydrogen produced by the different metals.
- Produce a graph of the results, which will allow the company to predict the volume of hydrogen generated from a specific mass of the different metals.
- Determine the chemistry of the reaction process for the different metals.
- Write a report to the company stating your findings and recommendations.

PROCEDURES

Your company will provide you with the materials and equipment needed to set up the apparatus shown in Figure 1 for your studies of hydrogen gas production from the different metals. Your supervisor will ask different teams to investigate a different metal (Table 1) and share the results. Based on the combined results, you are to determine the mass (g) of each metal needed to produce the same volume of hydrogen gas.

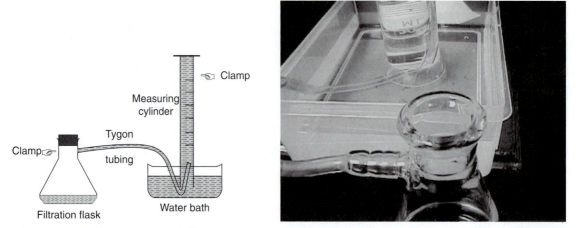

Figure 1 Hydrogen Gas Collection Apparatus

Table 1 Team Metals and Acid for $H_2(g)$ Production Studies

Team	Metal*	Acid
1	Zn	4 M HCl
2	Al	4 M HCl
3	Mg	1 M HCl**

* ≅ 1 g of the metals will be provided in the form of Mg ribbon, Al foil, and mossy zinc. Use sandpaper or a synthetic scrub sponge to clean and remove any coating from the Mg and Al.
** Note that Mg is to be tested with 1 M (rather than 4 M) of HCl.

PART I— MEASUREMENT OF HYDROGEN GAS

Caution: Hydrogen gas is explosive in the presence of air and an open flame!
No open flames are allowed while conducting this experiment!

1. What do you expect will happen to the volume of evolved hydrogen gas as the mass of metal is increased while acid concentration and volume are constant? Record your hypothesis.

   Hypothesis

2. Assemble the apparatus in Figure 1 for conducting your hydrogen gas study. Obtain and fill a large (250-mL) graduated cylinder with water and seal the top of the cylinder with parafilm. Make sure there is no air trapped at the top of the graduated cylinder. Fill a large container half-full with water to form a water bath. While holding the parafilm with one hand—so that no water can escape from the

cylinder—invert the cylinder and lower it into the water bath. When the top of the cylinder is below the level of the water in the water bath, remove the parafilm. Insert the open end of the tube shown in Figure 1 into the mouth of the graduated cylinder and hold the cylinder in position with a clamp.

3. Obtain a sample of your assigned metal that weighs about 0.02 grams. Record the exact mass in Table 2.

4. Use a graduated cylinder to obtain 25 mL of hydrochloric acid and add this acid to the flask.

> *Caution:* Do not allow acid to contact your skin!
> If contact occurs, wash the area profusely with cold water.
> Notify your instructor of any spills or body contact with HCl.

5. Add the metal to the flask as indicated below and then *immediately seal the top of the flask with the rubber stopper so that no gas is lost.*

   Mg ribbon or Al foil: Refer to Figure 2. Fold the metal and while carefully inclining the flask, place the folded metal in the neck of the flask. Immediately insert the stopper in the flask before the metal comes in contact with the acid.

   Mossy Zn: Add the metal quickly to the acid-containing flask and insert the stopper.

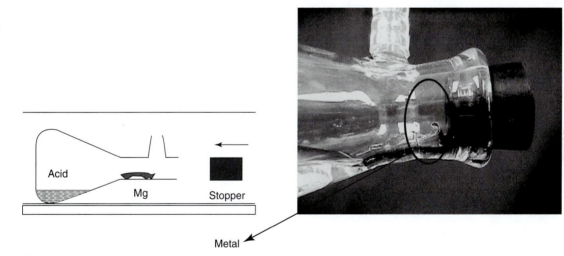

Acid   Mg   Stopper

Metal

Figure 2 Flask with Metal

6. Straighten the flask while carefully holding it (Figure 1) and clamp it loosely. Gently swirl the flask—as needed—until all of the metal has reacted. Read and record in Table 2 the volume of hydrogen gas that collects in the inverted graduated cylinder.

7. Repeat your study using a total of 6–8 samples of different known metal mass (0.02–0.20 grams).

Table 2 Mass of Metal versus Volume of Hydrogen Gas

Sample	Mass of _____ Metal (g)	Volume of Hydrogen (mL)

Data Analysis and Implications (Part I)

1. Examine your team results for any pattern regarding mass of metal consumed versus volume of hydrogen evolved. What do the results indicate?

2. Do the team results support or refute your hypothesis? Why or why not?

Your task is to determine and compare the mass (g) of each metal (Al, Mg, and Zn) needed to produce the same volume of hydrogen gas. In order to meet this objective you need to collect the results for the different metals tested by the different teams. Design a table (Table 3) for recording the compiled data from the different research teams. Make sure to include columns for the tested mass and evolved hydrogen gas volume for each of the different tested metals.

COMPILED TEAM DATA

Table 3 Mass (g) of Al, Mg, and Zn versus Gas Volume (mL)

Data Analysis and Implications (Part II)

3. Use the compiled team data to create a graph for determining and comparing results for the different metals. Plot the volume (mL) of gas evolved versus the mass (g) of the metal consumed for each metal on a single graph. Attach the graph to your report.

   a. Based on the graph of the combined team data, determine the mass of each tested metal required to produce 50, 75, and 100 mL of gas. Record your results in Table 4.

   Table 4 Mass of Metals Required for a Given Volume of Hydrogen Gas

Gas Volume	Mg (g)	Zn (g)	Al (g)
50 mL			
75 mL			
100 mL			

   b. Will the same mass of each metal produce the same volume of hydrogen gas?

4. Zinc metal is currently used for gas production. Compare the ratio by mass of Mg and Al to that of Zn for preparing the same volume of hydrogen gas. Use the data recorded in Table 3 for your calculations. Record your calculated results in Table 5.

   Table 5 Ratio by Mass of Metals Required for a Given Volume of Hydrogen Gas

Gas Volume	Mass Ratio Zn/Mg	Mass Ratio Zn/Al
50 mL		
75 mL		
100 mL		
Average		

Data Analysis and Implications (Part II continued)

a. The ratio of the atomic masses of Zn compared to Mg is 2.699. Does there appear to be a relationship between the mass of Zn and Mg metals required to prepare a volume of hydrogen gas and the ratio of the atomic masses of the metals? Refer to the data to support your answer.

b. Determine the ratio of the atomic mass of Zn to Al. Does there appear to be a relationship between the masses of these metals required to prepare a volume of hydrogen gas and the ratio of the atomic masses of the metals? Provide data examples to support your answer.

5. Are the same number of moles (mass/atomic mass) of each metal needed to generate the same volume of hydrogen gas? Use the mass data recorded in Table 4 to determine the ratio of mass to atomic mass of the different metals needed to prepare the same volumes of hydrogen gas. Record your results in table form (Table 6) and on a single graph (attach).

Table 6 Ratio of Metal Mass to Atomic Mass Required for a Given Volume of Gas

Gas Volume	Mass/At. Mass Mg	Mass/At. Mass Zn	Mass/At. Mass Al
50 mL			
75 mL			
100 mL			

Data Analysis and Implications (Part II continued)

a. Zinc is currently used for gas production. Are the number of moles of Mg and Al metal needed to produce a given volume of hydrogen gas the same as that of Zn metal? Explain.

b. The chemistry of hydrogen gas production for zinc metal is:

$$Zn\ (s) + 2H^+(aq) \rightarrow H_2\ (g) + Zn^{2+}(aq)$$

Provide balanced equations below for the chemistry of gas production for Mg and Al.

c. Use the results in Table 6 to: 1. calculate the ratio of moles (mass/at. mass) of Zn to moles of Al needed to produce the same volume of gas. 2. Based on the stoichiometry of your balanced equations, what is the ratio of moles of Zn compared to moles of Al needed to produce the same volume of gas? 3. Does there appear to be any relationship between the ratio calculated from the data and from reaction stoichiometry?

Experiment 2–1 Report

Provide a brief report to the company summarizing your findings and recommendations regarding the questions that follow. Make sure your report *refers to the compiled experimental data from the different teams.*

1. Based solely on experiment results, is magnesium or aluminum a reasonable alternative to zinc for producing a given volume of hydrogen gas?

2. Based on both the experimental results and the current value of the metals, is magnesium or aluminum a reasonable alternative to zinc? Search the Web for the updated prices of the metals or assume that the current values of the metals are $3.70/100 gram Zn, $3.70/100 gram Mg, and $1.80/100 gram Al.

1. Predict the mass (g) of magnesium that would be needed to generate 500 mL of hydrogen gas at room temperature and atmospheric pressure based on the compiled experiment data of the different teams.

$$Mg\,(s) + 2\,HCl\,(aq) \rightarrow Mg^{2+}(aq) + 2\,Cl^-\,(aq) + H_2\,(g)$$

2. Determine whether an unknown metal is Mg, Al, or Zn if 1.00 g of this metal reacts with excess acid according to the following equation to produce 374 mL of H_2 gas at room temperature and atmospheric pressure.

$$M\,(s) + 2\,H^+(aq) \rightarrow M^{2+}(aq) + H_2\,(g)$$

3. Calcium tablets are available in varying chemical forms as a dietary supplement. Compare the mass of the varying forms of tablets required to furnish 500 mg Ca^{2+}.

   a. calcium carbonate, $CaCO_3$
   b. calcium citrate, $Ca_3(C_6H_5O_7)_2$
   c. calcium lactate, $Ca(C_3H_5O_3)_2$

4. Explain why it takes 8 grams of oxygen per gram of hydrogen to form water.

$$2\,H_2\,(g) + O_2\,(g) \rightarrow 2\,H_2O\,(g)$$

5. How much oxygen must you inhale to consume 10.0 grams of the sucrose in cane sugar?

$$C_{12}H_{22}O_{11}\,(aq) + 12\,O_2\,(g) \rightarrow 12\,CO_2\,(g) + 11\,H_2O\,(l)$$

6. Equal volumes of oxygen and an unknown gas contain the same number of molecules and weigh 3.00 g and 7.50 g, respectively. Which of the following is the unknown gas?

   a. CO   b. CO_2 c. NO   d. NO_2 e. SO_2 f. SO_3

How Much Is Too Much?

It makes good "business sense" to plan the production of a product so as to keep cost and waste at a minimum. A local company plans to use some of their available magnesium and sulfuric acid or hydrochloric acid to prepare large quantities of hydrogen gas for use in their various processes. They have asked for your help in determining how to optimize the chemical reaction used for gas production. You are to determine experimentally the ratio of magnesium mass to acid concentration that will maximize the amount of hydrogen gas and minimize the amount of waste. You are also to determine, based both on experiment results and current acid price, whether sulfuric and hydrochloric acids are reasonable alternatives for gas production.

- Explore the effect of acid concentration on hydrogen gas yield.
- Produce a graph depicting the relationship of volume of hydrogen gas produced relative to acid concentration and magnesium mass.
- Determine the ratio of magnesium mass to acid concentration that maximizes gas yield.
- Write a report to the company in a scientific manner indicating how to maximize product yield and minimize waste.

The company will provide you with the materials and equipment needed to set up the apparatus shown in Figure 1 for your studies of hydrogen gas production. In order to produce a timely report, your supervisor will ask teams to use either sulfuric or hydrochloric acid for all research, as shown in Table 1. Based on the combined results of the different teams, you are to determine the ratio of magnesium mass to acid concentration that will maximize the amount of hydrogen gas.

> *Caution:* Hydrogen gas is explosive in the presence of air and an open flame!
> No open flames are allowed while conducting this experiment!

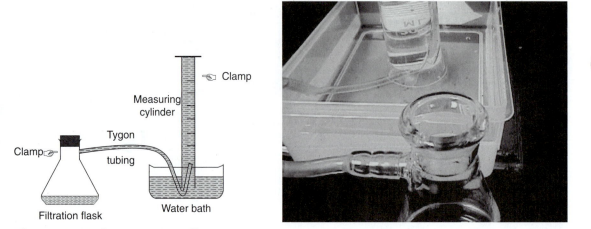

Figure 1 Hydrogen Gas Collection Apparatus

Table 1 Team Metal Mass and Acid for H_2 (g) Production Studies

Team	Metal*	Acid**
1	Mg (0.05g)	HCl
2	Mg (0.05g)	H_2SO_4
3	Mg (0.05g)	HCl
4	Mg (0.05g)	H_2SO_4

* Provided as Mg ribbon.
** Provided at a concentration of 1.0 M HCl and 1.0 M H_2SO_4.

1. What do you expect will happen to the volume of evolved hydrogen gas as acid concentration is increased from 0.10 M to 1.0 M while metal mass is constant? Record your hypothesis.

   Hypothesis

2. Assemble the apparatus in Figure 1 for conducting your hydrogen gas study. Obtain and fill a large (250 mL) graduated cylinder with water and seal the top of the cylinder with parafilm. Make sure there is no air trapped at the top of the graduated cylinder. Fill a large container half-full with water to form a water bath. While holding the parafilm with one hand—so that no water can escape from the cylinder—invert the cylinder and lower it into the water bath. When the top of the cylinder is below the level of the water in the water bath remove the parafilm. Insert the open end of the tube shown in Figure 1 into the mouth of the graduated cylinder and hold the cylinder in position with a clamp.

3. Read the guidelines and information that follow and then prepare and weigh your samples of magnesium and prepare your samples of assigned acid at different concentrations. Make sure to record your data below. *Note:* Different team results can only be shared and are only valid if all teams follow identical guidelines so that experiment conditions are the same.

Caution: HCl and H_2SO_4 cause burns! Avoid eye and skin contact! Wear goggles!

GUIDELINES

- Clean the Mg pieces before measuring and recording the mass of a strip. Use sandpaper or a synthetic scrub sponge to remove any coating.
- Measure the mass of Mg pieces weighing about 0.05 g ± .01 g.
- Use 5 mL of (HCl or H_2SO_4) acid at four different acid concentrations with a minimum concentration of 0.10 M and a maximum concentration of 1.0 M.
- Dilute the 1.0 M acid to prepare acid solutions of different concentrations.
- Use a pipet or buret to deliver the desired volume of the 1.0 M acid (HCl or H_2SO_4) into a *clean* 10 mL volumetric flask. Add water to the 10 mL calibration line to obtain 10 mL of acid solution of a desired concentration.
- Measure and record the volume of hydrogen produced in mLs.

TEAM DATA

Mg mass (g)

Preparation of different concentrations (M) of acid _____.

4. Use a small graduated cylinder to add 5 mL of your assigned acid (HCl or H_2SO_4) of known concentration to the reaction flask ($\cong$ 125 mL).

5. While carefully inclining the flask (Figure 2), place the folded piece of magnesium of known exact mass in the neck of the flask and insert the stopper *before* the metal comes in contact with the acid. *Note:* The magnesium metal may be cut into a few pieces or inserted as one piece. Place the flask upright and immediately seal the top of this flask with the rubber stopper so that no gas is lost. Gently swirl the flask, as needed, until all of the metal has reacted. Read the volume (mL) of hydrogen gas that collects in the inverted graduated cylinder and enter your team data in Table 2.

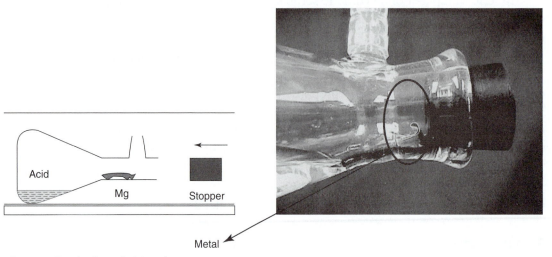

Figure 2 Flask with Metal

6. Repeat your study using different concentrations of your assigned acid.

Table 2 Team Data: Mg (g) and _____ Acid (M) versus Volume (mL) of H_2

Sample	Mg (g)	Acid (mL)	Acid Molarity	H_2 (mL)
1				
2				
3				
4				

7. Collect and compile the different team results using a computer or Table 3.

Table 3 Mass of Mg, Acid Concentration, and Volume of H$_2$

Mg (g)	H$_2$SO$_4$ (mL)	H$_2$SO$_4$ (M)	H$_2$ (mL)

Mg (g)	HCl (mL)	HCl (M)	H$_2$ (mL)

Optional Extension

How would results compare if the study were done using a constant concentration of acid and different amounts of Mg? For example, study the impact of magnesium mass (0.05–0.35 g) on the volume of evolved hydrogen gas using acid at a constant concentration (2 M).

Data Analysis and Implications

Submit the information below to your supervisor.

1. Plot the combined data of all teams (volume of gas evolved versus acid concentration for each of the acids) on a single piece of graph paper. Attach the graph and answer questions (a and b).

    a. What are the implications regarding volume of hydrogen gas (H_2) evolved and concentration of acid?

    b. Based on the plot, which reactant disappears first (i.e. which is the limiting reactant that is used up completely) and which reactant is present in excess form?

2. Calculate the number of moles of H_2 generated per 0.5-g sample (2.06×10^{-2} moles) of consumed magnesium for each acid. The number of moles of H_2 can be calculated using the relationship, $PV = nRT$ (the Ideal Gas Equation). What is the value of the following ratio for each acid (HCl or H_2SO_4) based on the graph?

$$\frac{\text{Mole/s of } H_2}{\text{Mole of magnesium}} = \underline{\hspace{2cm}} \text{ for HCl}$$

$$\frac{\text{Mole/s of } H_2}{\text{Mole of magnesium}} = \underline{\hspace{2cm}} \text{ for } H_2SO_4$$

● **Experiment 2–2 Report**

Provide a brief report to the company summarizing your findings and recommendations. Make sure your report *refers to your data* and answers the following questions:

1. What ratio of magnesium mass to acid (HCl and H_2SO_4) concentration will maximize the amount of hydrogen gas and minimize waste?

2. Does the identity of the acid matter relative to hydrogen gas yield? Why or why not?

3. Based on both the experiment results and the current price of sulfuric acid and hydrochloric acid, is either acid a reasonable alternative for hydrogen gas production? Search the Web for updated prices of the acids or use this price listing: 2.5 liters of sulfuric acid = $20.00 and 2.5 liters of hydrochloric acid = $12.00.

1. How many pizzas can you make if you use 750 grams of dough, 250 grams of cheese, and 35 slices of pepperoni on each pizza? Assume that you have 3500 grams of dough, 1500 grams of cheese, and 250 slices of pepperoni. What is the "limiting reagent" in this exercise?

2. How many H_2O molecules can you make from 500 H_2 molecules and 500 O_2 molecules?

3. What is the molarity of a solution of hydrochloric acid, HCl, if 5.00 mL reacts completely with 20.00 mL of NaOH to produce water and NaCl? The NaOH is known to contain 2 mmol of hydroxide ions (OH^-) per milliliter.

4. What is the molarity of a solution of sulfuric acid, H_2SO_4, if 5.00 mL reacts completely with 20.00 mL of NaOH to form water and Na_2SO_4? The NaOH is known to contain 2 mmol of hydroxide ions (OH^-) per milliliter.

5. When cane sugar reacts with oxygen in living systems, carbon dioxide and water are produced, and a great deal of energy is liberated:

$$C_{12}H_{22}O_{11}\ (s) + 11\ O_2\ (g) \rightarrow 12\ CO_2\ (g) + 11\ H_2O\ (g)$$

What weight of carbon dioxide can be produced from the reaction of 10.0 grams of cane sugar with 10.0 grams of oxygen?

6. J. J. Berzelius found that 11.56 grams of lead sulfide, PbS, were formed when 10.0 grams of lead reacted with 1.56 grams of sulfur, when 10.0 grams of lead reacted with 3.00 grams of sulfur, and when 18.0 grams of lead reacted with 1.56 grams of sulfur. Explain his observations in terms of the concept of limiting reagent.

Experiment 3–1

What Makes a Solution Colored or Colorless?

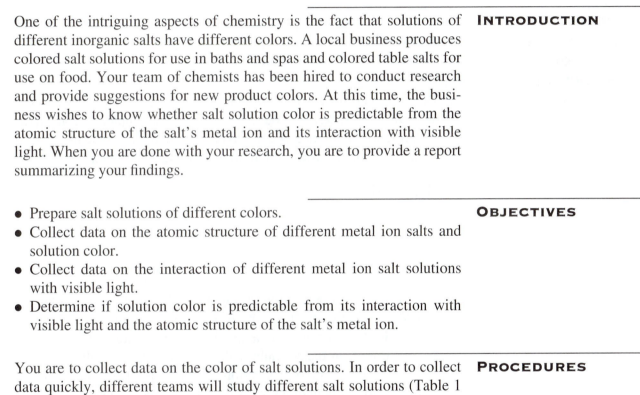

One of the intriguing aspects of chemistry is the fact that solutions of different inorganic salts have different colors. A local business produces colored salt solutions for use in baths and spas and colored table salts for use on food. Your team of chemists has been hired to conduct research and provide suggestions for new product colors. At this time, the business wishes to know whether salt solution color is predictable from the atomic structure of the salt's metal ion and its interaction with visible light. When you are done with your research, you are to provide a report summarizing your findings.

INTRODUCTION

- Prepare salt solutions of different colors.
- Collect data on the atomic structure of different metal ion salts and solution color.
- Collect data on the interaction of different metal ion salt solutions with visible light.
- Determine if solution color is predictable from its interaction with visible light and the atomic structure of the salt's metal ion.

OBJECTIVES

You are to collect data on the color of salt solutions. In order to collect data quickly, different teams will study different salt solutions (Table 1 or as indicated by your supervisor) and share the results.

PROCEDURES

Table 1 Team Assignments: 0.10 M Salt Solutions[*]

Team 1	$Al(NO_3)_3$	(1)	$Cr(NO_3)_3$	(4)
Team 2	$Cd(NO_3)_2$	(2)	$Co(NO_3)_2$	(5)
Team 3	$Ca(NO_3)_2$	(3)	$Cu(NO_3)_2$	(6)
Team 4	$SnCl_2$ in HCl	(11)	$Ni(NO_3)_2$	(8)
Team 5	$NaNO_3$	(10)	$Fe(NO_3)_3$ in HNO_3	(7)
Team 6	$Zn(NO_3)_2$	(12)	$Ni(NO_3)_2$ in NH_3	(9)

[*] Refer to the indicated (number) in Table 2 for information on the assigned salt solution.

PART I— PREPARATION AND COLOR OF SOLUTIONS

Each team member will prepare one assigned salt solution using the procedure below. When you are done, store your solution for studying its interactions with visible light (Part III).

1. Does the color of a salt solution relate to the atomic structure of the salt's metal ion? Record your hypothesis. For example, do you expect the presence or absence of color to be predictable from the metal ion's ionic radius or electron configuration or some other factor?

   Hypothesis

2. Record the formula of your assigned salt. Typical salt formulas are provided in Table 2. Check the formula on the reagent bottle since hydrated salts often are available with different mole ratios of water to salt. Based on the formula, calculate the mass of salt needed to prepare 25 mL of 0.10 M solution. Work individually, but feel free to check your calculations with your team since everyone depends on everyone else's work.

3. Measure out and record the mass of salt to the nearest 0.01g. *Note*: It is difficult to gauge the amount of salt needed for an accurate precalculated mass. All that you need to do is to measure out an amount that is close to your calculated value. Then, calculate the *actual* concentration (M) of your sample based on the actual mass of salt that you use.

4. Place a funnel into the 25-mL volumetric flask and transfer the salt sample into that flask. Add *a little* of the appropriate solvent. (*Note:* All solutions are prepared with water except for those cited under Table 2.) Swirl the flask until the solid dissolves completely. Add more solvent until the bottom of the meniscus just touches the flask's calibration mark when your eye is level with it. Mix the contents.

5. Record the color of your prepared solution. Compile the different team solution color data using a computer or use Table 2 provided. Label and store your prepared solution for use in Part III.

Table 2 0.10 M Salt Solution Color

	Salt	Typical Reagent Formula	Solvent	Solution Cation Species	Solution color* (V, B, G, Y, O, R) or colorless?
1.	Aluminum nitrate	$Al(NO_3)_3 \bullet 9H_2O$	H_2O	$Al(H_2O)_6^{3+}$	
2.	Cadmium nitrate	$Cd(NO_3)_2 \bullet 4H_2O$	H_2O	$Cd(H_2O)_6^{2+}$	
3.	Calcium nitrate	$Ca(NO_3)_2 \bullet 4H_2O$	H_2O	$Ca(H_2O)_6^{2+}$	
4.	Chromium(III) nitrate	$Cr(NO_3)_3 \bullet 9H_2O$	H_2O	$Cr(H_2O)_6^{3+}$	
5.	Cobalt(II) nitrate	$Co(NO_3)_2 \bullet 6H_2O$	H_2O	$Co(H_2O)_6^{2+}$	
6.	Copper(II) nitrate	$Cu(NO_3)_2 \bullet 2.5H_2O$	H_2O	$Cu(H_2O)_4^{2+}$	
7.	Iron(III) nitrate	$Fe(NO_3)_3 \bullet 9H_2O$	HNO_3	$Fe(H_2O)_x^{3+}$	
8.	Nickel nitrate	$Ni(NO_3)_2 \bullet 6H_2O$	H_2O	$Ni(H_2O)_6^{2+}$	
9.	Nickel nitrate	$Ni(NO_3)_2 \bullet 6H_2O$	NH_3	$Ni(NH_3)_6^{2+}$	
10.	Sodium nitrate	$NaNO_3$	H_2O	$Na(H_2O)_6^{2+}$	
11.	Tin(II) chloride	$SnCl_2 \bullet 2H_2O$	HCl	$Sn(H_2O)_6^{2+}$	
12.	Zinc nitrate	$Zn(NO_3)_2 \bullet 6H_2O$	H_2O	$Zn(H_2O)_6^{2+}$	

* **V** = violet, **B** = blue, **G** = green, **Y** = yellow, **O** = orange, **R** = red.

All solutions are prepared with water except for those noted below:

7. Iron(III) nitrate. Use 0.10 M HNO_3 instead of water.

9. Nickel nitrate: Transfer 10 mL of 0.10 M $Ni(NO_3)_2$ and add 1.0 M of NH_3 to the 25-mL calibration line of the volumetric flask.

11. Tin(II) chloride. Use 10% HCl instead of water.

Optional Extension   Dissolve a pinch of nickel nitrate, cobalt(II), nitrate and copper(II) nitrate in water and observe the solution colors. Repeat using a solvent other than water. Is solution color solvent-dependent?

Data Analysis and Implications (Part I)

1. Record the observable color for each salt solution in the position of the salt's metal ion element in the periodic table (Figure 1).

1 IA	2 IIA	3 IIIB	4 IVB	5 VB	6 VIB	7 VIIB	8 VIIIB	9 ←→	10 VIIIB	11 IB	12 IIB	13 IIIA	14 IVA	15 VA
1 H Hydrogen														
3 Li Lithium	4 Be Beryllium													
11 Na Sodium	12 Mg Magnesium											13 Al Aluminum		
19 K Potassium	20 Ca Calcium	21 Sc Scandium	22 Ti Titanium	23 V Vanadium	24 Cr Chromium	25 Mn Manganese	26 Fe Iron	27 Co Cobalt	28 Ni Nickel	29 Cu Copper	30 Zn Zinc	31 Ga Gallium	32 Ge Germanium	
37 Rb Rubidium	38 Sr Strontium	39 Y Yttrium	40 Zr Zirconium	41 Nb Niobium	42 Mo Molybdenum	43 Tc Technetium	44 Ru Ruthenium	45 Rh Rhodium	46 Pd Palladium	47 Ag Silver	48 Cd Cadmium	49 In Indium	50 Sn Tin	51 Sb Antimony

Figure 1 Salt Solution Color vs. Placement of the Metal Ion's Element in the Periodic Table

2. Look up the color of the salts L_1NO_3, KNO_3, $Mg(NO_3)_2$ and $Ga(NO_3)_3$ in the CRC Handbook of Chemistry and Physics or another reference text. Record the color for solutions of these salts in Figure 1. Note that any white salt will be colorless in aqueous solution.

3. Answer questions a and b based on the data in Figure 1:

   a. Does there appear to be a relationship between the presence and absence of color in a salt solution and the position of the element in the periodic table? Record and support your decision with specific examples from the collected team data.

   b. Can you predict whether an untested metal ion solution will be colored or colorless? For example, based on the data entered into Figure, do you expect a solution of rubidium nitrate to be colored or colorless? Why or why not?

Your goal is to collect data on the atomic structure of different metal ions to answer the question: Is the presence or absence of solution color linked to the structure of the salt's metal ion?

1. Figure 2 gives the valence shell electron configurations of some elements. Record the valence shell electron configurations of the metal ions tested in Part I and the researched metal ions (L_1^+, K^+, Mg^{2+}, and Ga^{3+}) in Figure 3. Note that the electron configuration of a metal ion is different from its metal atom and reflects the electrons lost by the metal to form the metal ion. For example, the electron configuration of Zn is $[Ar]4s^23d^{10}$, but Zn^{2+} is $[Ar]3d^{10}$.

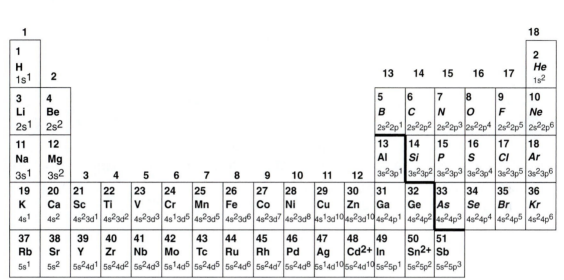

Figure 2 Valence Shell Electron Configurations of Some Elements

Figure 3 Electron Configuration of Tested Metal Ions

Record any observable pattern(s) in electron configurations and placement of metal ions in the periodic table.

2. Figure 4 gives the ionic radii of some common metal ions.

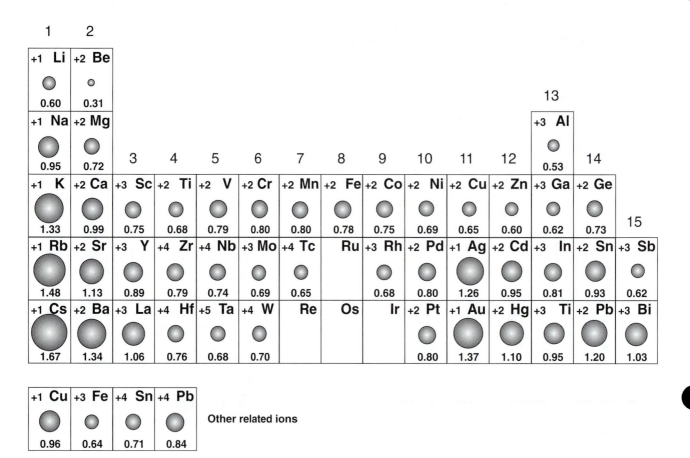

Figure 4 Ionic Radii (picometers) of Some Common Metal Ions

Record any pattern in ionic radii size and placement of metal ions in the periodic table.

Data Analysis and Implications (Part II)

3. Does there appear to be a relationship between the presence or absence of solution color and the electronic configuration of the metal ions? Support your conclusion using specific examples of metal ion electron configuration (Figure 3) and the compiled team data on solution color (Table 2).

4. Is there a relationship between the presence or absence of solution color and the ionic radius of the metal ion? Support your decision using specific examples from the compiled team data on solution color (Table 2) and metal ion radii (Figure 4).

You will now collect data on the interaction of colorless and colored salt solutions with visible light. Teams will investigate the interaction of the prepared (Part 1) assigned salt solutions (Table 1) with visible light. Share the resulting data to determine an explanation for the observable color of samples.

Your research begins with a determination of the colors of the wavelengths comprising visible light using a spectrophotometer or colorimeter. Place a cuvette holding a piece of plain white paper in the sample chamber. Set the wavelength at 420 nm and observe the color of the light reflected from the paper. Record the observed color in Table 3 at 20-nm intervals from 420 to 680 nm.

Part IIIA—Wavelength Color

Spectronic 20 or Colorimeter

Open the sample chamber and look inside to see the color of the light reflected from the paper.

UV-Vis Spectronic Genesys 5

Place the cuvette in the first sample chamber so that the paper is facing the light slit. At the main menu select #1 (ABS/%T/CONC) by pressing the soft key numeral 1. At the top of the soft keypad, press the arrow above **Go to WL.** Using the soft numeral pad, enter the wavelength 420 nm. Make sure the spectrophotometer is looking at the first sample chamber. By opening the large access lid you should be able to see the wavelength color on the white piece of paper. Leave the lid open and change to your desired wavelength with the **Go to WL** command.

Table 3 Visible Light Wavelength Colors

λ (nm)	Color	λ (nm)	Color
420		560	
440		580	
460		600	
480		620	
500		640	
520		660	
540		680	

**Part IIIB—Color and
Visible Light Interaction**

1. Record your hypothesis regarding the relationship between the observed color of a solution and its interaction with visible light. For example, will the absorbance of light for a green solution be maximum or minimum at wavelengths of light that are green?

   Hypothesis

2. Measure the absorbance and transmission of your assigned 0.10 M solution using a spectrophotometer or colorimeter. Record the absorbance and transmission data from wavelengths 420 nm to 680 nm in Table 4 using the guidelines.

   GUIDELINES

   - Use two cuvettes, one to contain your reference blank (distilled water or solvent used to prepare the solution) and the other to contain the solution sample.
   - Make sure your cuvette is filled high enough (about 2/3 full) so that the visible light beam will pass through the solution rather than through air.
   - Make sure the cuvette exterior is clean and dry.
   - Record absorbance and transmission readings at 20-nm intervals from 420 nm to 680 nm.
   - In regions of high absorbance, take readings at 10 nm or smaller intervals to obtain a smoother curve.
   - Calibrate the spectrophotometer every time you change the wavelength.

Table 4 Absorbance and Transmission Data for _____

λ	Absorbance	Transmission	λ	Absorbance	Transmission

3. Plot an absorption spectrum graph (a graph of absorbance as a function of λ). On the spectrum, record the color, concentration, and identity of the solution.

4. Use your spectrum data and your color vs. wavelength data (Part IIIA) to determine the wavelength and wavelength color at which each team-assigned compound absorbs the most light. This wavelength is called the λ_{max} of your compound. Similarly, determine the wavelength and wavelength color at which your compound transmits the most light. Record your data in Table 5. If your sample has several absorbance peaks, record data for the secondary peaks in Table 6.

TEAM DATA

Table 5 Primary Absorbance and Transmission Wavelength Color for _____

Salt Solution	Abs. λ_{max}	Abs. λ_{max} Color	Trans. λ_{max}	Trans. λ_{max} Color

Table 6 Secondary Absorbance and Transmission Wavelength Color for _____

Salt Solution	Abs. λ_{max}	Abs. λ_{max} Color	Trans. λ_{max}	Trans. λ_{max} Color

5. Do your results support your hypothesis about observed solution color and visible light interaction? Why or why not?

6. Collect the different team data on solution color and light interaction using a computer or Table 7 provided below.

COMPILED TEAM DATA

Table 7 Solution Color versus Abs. λ_{max} Color and Trans. λ_{max} Color

	Salt Solution	Solvent	Visible Color[*] V B G Y O R	Abs. λ_{max} Color/s	Trans. λ_{max} Color/s
1.	Aluminum nitrate	H_2O			
2.	Cadmium nitrate	H_2O			
3.	Calcium nitrate	H_2O			
4.	Chromium(III) nitrate	H_2O			
5.	Cobalt(II) nitrate	H_2O			
6.	Copper(II) nitrate	H_2O			
7.	Iron(III) nitrate	HNO_3			
8.	Nickel nitrate	H_2O			
9.	Nickel nitrate	NH_3			
10.	Sodium nitrate	H_2O			
11.	Tin(II) chloride	HCl			
12.	Zinc nitrate	H_2O			

[*] V = violet, B = blue, G = green, Y = yellow, O = orange, R = red.

Data Analysis and Implications (Part III)

5. What is the relationship, if any, between solution color and visible light interaction? For example, can solution color be predicted from the Abs λ_{max} or Trans λ_{max} color(s)? Support your conclusion by citing specific examples from the compiled team data in Table 7.

Experiment 3–1 Report

Summarize your findings regarding salt solution color, structure of the salt's metal ion, and visible light interactions. Support your conclusions with specific examples from the compiled data of the different teams.

1. Will aqueous solutions of the following salts be colored or colorless? Titanium(II) chloride and beryllium chloride

2. Predict whether nitrate salt solutions containing metal ions with the following outer shell electron configurations will be colored or colorless: $[Ar]3d^2$ and $[Kr]4d^{10}$.

3. The graph below shows the absorption spectra for an indicator in acidic and basic solutions. What color is the indicator in acidic solution and in basic solution?

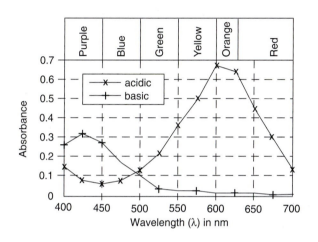

4. Identify the visible absorption spectrum of an aqueous solution of sugar from the spectra (**A** or **B** or **C**) given below.

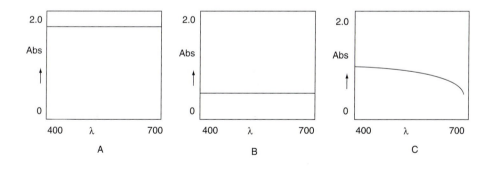

5. You spray paint the letters CHEMISTRY in blue paint on snow. After a while the sun comes out and the snow under the blue paint melts, putting the letters in the word CHEMISTRY in relief. Why did the snow under (rather than around) the blue paint melt?

6. Search the World Wide Web for answers to the following color-related questions:

a. Colored table salt would look great on food and you might use less! Are colored salts available for purchase? What gives the salt its color?

b. Green dye lasers are used in eye surgery to prevent eye damage during surgery. Why?

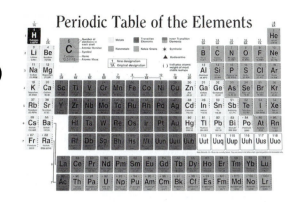

Periodic Table of the Elements

Are There Property Patterns?

Chemists often look for patterns in the properties of different tested materials in order to predict the properties of untested samples. For example, it seems plausible to imagine that substances containing "lighter" molecules or atoms will melt at lower temperatures than substances made up of "heavier" atoms or molecules. The chemist will test this idea by determining whether a set of samples exhibit a mass and melting-point pattern and, based on results, will either refine or discard the idea. In this investigation, your team will test ideas relating atomic or molecular structure and a property. Your task is to identify groups of substances where the correlation works and explore possible reasons for its failure when it doesn't.

- Use the Internet and chemical reference books to obtain property data about elements and compounds.
- Plot properties against atomic structures of elements and note any patterns including periodic trends.
- Experimentally test the property and reactivity of compounds and look for structure and property relationships.
- Use atomic and molecular structures of elements or compounds to explain regularities or discontinuities in properties.

This inquiry asks you to use the Internet, chemical reference books (Part I), and experimental tests (Part II) to obtain and plot property data about samples. Part I may be conducted in lab if computers are available or done as a pre-lab prior to conducting Part II.

Internet Web sites such as **webelements** at www.webelements. com are available that allow one to determine or plot properties of the elements against atomic structure. Such data can alternatively be found in the *CRC Handbook of Chemistry and Physics* and then plotted by hand or with the aid of a software program. The procedure below refers to the www.webelements.com site, but your instructor may refer you to other Internet Web sites.

Read the guidelines below before starting.

GUIDELINES

- Start your inquiry by recording your hypothesis regarding a possible relationship between the property and atomic number. The hypothesis needn't be correct in fact. A hypothesis is a reasonable statement that can be tested and disproved by experiment.
- When you have completed the plot, record any noticeable trends:
    1. Is there a general trend in the property as atomic number increases?
    2. Is there an observable property pattern across a period of elements?
    3. Is there a group of elements that exhibit a particularly high or low property value?

Access the Web site www.webelements.com as indicated by your instructor. On the main page, you will observe a periodic table. To obtain property data and graphical plots, you will need to click on any element in the periodic table and follow the directions indicated below. After completing a property investigation, return to the home page of webelements by clicking on "Pro Home." Then click on any element in the periodic table, and proceed with the next investigation.

Part IA—Melting Point Patterns

1. Under the heading **Elemental properties** located on the left side of the screen, click on **Thermal properties**. Under the heading **Temperatures** click on **Melting point** and record its definition.

2. Do you expect to find a general trend in the melting point of elements as atomic number increases? Record your hypothesis.

    Hypothesis

3. View a melting point (*y*-axis) versus atomic number (*x*-axis) plot. For example, select **Bar** in the select box under **Full table charts**.

4. Record any noticeable trends in melting points as indicated in the above guidelines. Do the data support or contradict your hypothesis?

Part IB—Density Patterns

1. Under the heading **Elemental properties** located in the column on the left side of the page, click on **Bulk Properties**. Under the heading **Bulk Properties**, click on **Density of solid**. Record the definition of density.

2. Do you expect to find a general trend in the density of elements as atomic number increases? Record your hypothesis.

   Hypothesis

3. View a bar plot of density versus atomic number from among the choices found under **Full table charts**. You may view other types of plots such as a scatter plot, ball chart, or shaded table by clicking on the appropriate choice listed under the graph.

4. Record any noticeable trends in density and atomic number as indicated in the above guidelines. Do the data support or contradict your hypothesis?

Part IC—Atomic Radius Patterns

1. Under the heading **Electronic properties** in the left column, click on **Atom radii**. Record the definition of atomic radius.

2. Do you expect to observe a general trend in the atomic radius of elements as the atomic number of elements increases? Record your hypothesis.

   Hypothesis

3. View a bar plot of atomic radius versus atomic number. To do this, select **Bar** in the **View** box by **Atomic radius** and click **Go**. Record any noticeable radius patterns.

4. Select several groups under the heading **Group charts**. View the plots and look for patterns within the families of the main group elements as well as the transition metals.

5. Select several periods under the heading **Period charts**. View the plots and look for patterns within the periods of the main group elements as well as the transition metals.

1. Under the heading **Electronic properties** click on **Electronega-tivities**. Read and record the definition of **Pauling electronegativity**.

2. Do you expect to observe a general trend in the electronegativity of elements as atomic number increases? Record your hypothesis.

   Hypothesis

3. View a bar plot of electronegativity versus atomic number from among the choices found under **Full table charts**. Record any noticeable trends.

4. Select several periods under the heading **Period charts**. Record any electronegativity patterns within the periods of the main and transi-tion group elements such as periods 2, 3, and 4.

5. Select several groups under the heading **Group charts**. Look for electronegativity patterns within the families of the main group elements including the halogen family (group 17). Note that in Part II you will conduct a wet lab involving the halogen family.

Optional Extensions

1. Investigate any discontinuities in the way the property (atomic radius, density, melting point) changes with atomic number.

2. Compare the data obtained from the Internet site **webelements** with that available in the *CRC Handbook* and/or another Web site. Are the data similar or different?

Data Analysis and Implications (Part I)

1. Is there a general property (melting point, radius, electronegativity, density) trend of elements as the atomic number increases? Describe any pattern for:

   a. melting point

   b. density

   c. radius

   d. electronegativity

2. Is there a general trend in property of elements across a period of elements? Describe any pattern for:

   a. radius

   b. electronegativity

3. Explain any observable pattern in electronegativity across a period of elements based on atomic structure.

Your task is to identify the products of halogen and halide mixtures (Part IIB) based on your observations of the properties of halogens and halides (Part IIA). Your investigation starts with a study of the color and solubility of the halogens and halides in water and hexane.

Part II—Properties of Halogens and Halides

INFORMATION

Part IIA—Solubility and Color of the Halides and Halogens

- The halogens (chlorine = Cl_2, bromine = Br_2, and iodine = I_2) are molecular.
- The halides (chloride = Cl^-, bromide = Br^-, and iodide = I^-) are ions of the halogens.
- Silver and halide ions, $X^-(aq)$, react to form a solid silver halide precipitate: $Ag^+(aq) + X^-(aq) \rightarrow AgX(s)$

1. Transfer 2–3 mL of water to a test tube. Transfer 2–3 mL of hexane to a separate test tube. Slowly add the contents of the test tube containing water to the test tube containing hexane. Record your observations and conclusions regarding the solubility of hexane and water.

2. Test the solubility of the halides Cl^-, Br^-, and I^- in both water and hexane by adding a small amount (tip of spatula) of solid NaCl, NaBr, or NaI to separate small test tubes and then adding 2–3 mL of water. **Mix well** (agitate for 15–20 seconds). Record the color of each solution. Is each solid halide soluble or insoluble in water? (*Note:* Hold the tube and contents against a dark background so as to better observe the solid.) Save the labeled tubes and contents. Repeat the test using 2–3 mL of hexane instead of water. Is each solid halide soluble or insoluble in hexane? **Save the labeled tubes and contents** for step 3. Record your observations and conclusions.

3. Mix the Cl^- added to water with the Cl^- added to hexane and shake well. Repeat this procedure with the Br^- and I^- samples. Carefully add 1–2 drops of 0.1 M $AgNO_3(aq)$ to each mixture. Record your

conclusions regarding solubility and color of the halides in water and hexane.

> *Caution:* AgNO$_3$ will stain your skin and clothing!
> Avoid skin contact and wear gloves if available.

4. Test the solubility of I$_2$ crystals in water and hexane. Add a grain or two of the solid iodine to a small test tube and add 2–3 mL of water. Add a few mL of hexane to the water solution of I$_2$ and mix vigorously. Record your observations and conclusions. Compare your solubility and color results of the halogen I$_2$ with the results of its halide ion I$^-$ tested above in steps 2 and 3.

5. Obtain aqueous solutions of the halogens Cl$_2$, Br$_2$, and I$_2$. Record the color of each solution.

> *Caution:* Do not breathe chlorine, bromine, and iodine—work in the hood!
> Rinse any affected skin for 10 minutes under running water.
> Anyone with a history of asthma or bronchial disease should notify the instructor.

Transfer 0.5–1 mL (10–20 drops) of each halogen solution to separate tubes. Add 2–3 mL of hexane to each tube and **mix well**. Record your observations and conclusions regarding solubility and color in water vs. hexane. Record the color of the hexane layer after mixing as "color in hexane."

6. Compare your results with those of another team. If necessary, repeat some tests to resolve differences in results.

7. Share your team data and collect the different team data using Table 1 and Table 2 or a computer or as indicated by your instructor.

COMPILED TEAM DATA

Table 1 Solubility and Color of the Halides[*]

Team	Water			Hexane		
	Cl^-	Br^-	I^-	Cl^-	Br^-	I^-

[*] S = soluble; I = insoluble.
NC = no color, B = brown, P = pink, V = violet, YG = yellow-green, YO = yellow-orange.

Table 2 Solubility and Color of the Halogens[*]

Team	Water				Hexane			
	I_2 Solid	Cl_2 Water	Br_2 Water	I_2 Water	I_2 Solid	Cl_2 Water	Br_2 Water	I_2 Water

[*] S = soluble; I = insoluble.
NC = no color, B = brown, P = pink, V = violet, YG = yellow-green, YO = yellow-orange.

Part IIB—Reactivity of the Halides and Halogens

INFORMATION

If a halogen (X_2) reacts with halide ions (Y^-) of another halogen, then the reacting halogen (X_2) is considered more reactive than the product halogen (Y_2). Similarly, the reacting halide species (Y^-) is considered more reactive than the product halide (X^-). The symbol $>$ or $<$ is used to record comparative activity:

$$X_2 + 2Y^- \rightarrow Y_2 + 2X^- \quad \therefore \quad X_2 > Y_2 \ (\text{and } Y^- > X^-)$$

1. Based on your Part IA atomic structure investigations, formulate and record a hypothesis regarding the comparative reactivity of the halogens bromine, Br_2, chlorine, Cl_2, and iodine, I_2.

   Hypothesis

2. Add 10 drops of Br_2 water to 3 mL of 0.10 M NaCl. Add 2 mL of hexane and mix vigorously. After thorough mixing, record your observations of the hexane and aqueous layers in Table 3 and write a balanced net equation in Table 4 describing any net chemical reaction you observed. If you observed no reaction, write "no reaction."

3. Repeat step 2 so that you combine each halogen with each halide in the presence of hexane and determine whether a reaction occurs. When you are finished, you should have recorded observations for all the indicated combinations in Tables 3 and 4.

GUIDELINES

- Use small test tubes or vials for your investigations.
- Use $\cong 2$ mL (40 drops) of each halide (0.10 M NaCl, 0.1 M NaBr, or 0.1 M NaI).
- Add $\cong 2$ mL of hexane to each halide solution.
- Use $\cong 0.5$ mL (10 drops) of the halogen (in water).
- **Mix vigorously** (the hexane and water layer must make contact).
- Use a white background to better observe any change in solution color.

Table 3 Observations of Halogen + Halide Mixtures

	Layers	Cl^-	Br^-	I^-
+ Br_2	hexane			
	water			
+ Cl_2	hexane			
	water			
+ I_2	hexane			
	water			

Table 4 Halogen + Halide Reaction Equations

Halide + Halogen Mixture	Reaction? (yes or no)	Balanced Net Reaction Equation
$Cl^- + Br_2$		
$Br^- + Br_2$		
$I^- + Br_2$		
$Cl^- + Cl_2$		
$Br^- + Cl_2$		
$I^- + Cl_2$		
$Cl^- + I_2$		
$Br^- + I_2$		
$I^- + I_2$		

Date: _____ Name: _____ Class: _____

Data Analysis and Implications (Part II)

4. Complete Table 5 based on the collected data. Indicate whether a halogen and halide mixture reacted or did not react. Use the results to complete the following statements regarding halogen and halide reactivity. Compare the reactivity of the halogens Br_2, Cl_2, and I_2, based on the data.

Table 5 Reactivity of the Halogens

	Cl^-	Br^-	I^-	# Reactions Halogens
Cl_2				
Br_2				
I_2				
# Reactions Halides				

a. Reactivity of the halogens: _____ > _____ > _____

b. Reactivity of the halides: _____ > _____ > _____

5. Is there a relationship between the reactivity of the halogens and halides and their position in the periodic table? For example, can you predict what would occur if F_2 and At_2 were reacted with I^-? Would a reaction occur? Why or why not?

6. Organize the data collected in Part II in the form of a graph (attach) to depict the reactivity of the halogens vs. electronegativity. Indicate whether electronegativity differences can account for the observed differences in halogen reactivity.

Experiment 3–2 Report

Summarize the results of your studies of the relationship between properties of elements and compounds and their structure including periodic trends. Include graphs or data tables depicting property and structure relationships to support your conclusions.

1. On the basis of their positions in the periodic table, select the atom with the larger atomic radius in each of the pairs below:

   a. K and Fr
   b. Mg and Ba
   c. C and Sn
   d. Ar and Rn

2. A helium atom has a smaller atomic radius than the hydrogen atom. Explain why.

3. The most chemically reactive metal elements are soft metals with relatively low melting points. These metals are so reactive that they react spontaneously with water and/or substances in the atmosphere and therefore must be stored under oil. Where is this family of metals located in the periodic table?

4. Many expensive units of stereo and other electronic equipment use gold-coated parts and connectors because gold, Au, is less reactive and therefore has a higher resistance to oxidation than other metal parts. Can the reactivity of gold be explained based on its position in the periodic table?

5. Potassium, K, is more reactive than sodium, Na. Calcium, Ca, is less reactive than sodium. The comparative reactivity of these metals is often noted as: K > Na > Ca. Explain the reactivity differences of K, Na, and Ca.

6. Fluorine reacts with almost anything it touches. Toothpastes containing fluorides help maintain the toughness of tooth enamel, $Ca_{10}(PO_4)_6(OH)_2$, by replacing the hydroxide ion with fluoride ion. The resulting fluoride enamel is nonreactive. Use the data from this experiment to explain this observation.

Experiment 3–3

What Factors Affect Color Intensity?

This past summer unique organisms were found living in the local sewers. Local water and sewer treatment plant personnel investigated the situation. Not much was determined about the organisms, but it is known that their blood contains high concentrations of salt, a different salt for each species. One result of the presence of these organisms is colored sewer water. The treatment plant has asked your team of chemists to determine the factors that affect the intensity of salt solution color.

INTRODUCTION

OBJECTIVES

- Prepare salt solutions of different color intensity.
- Measure color intensity of each salt solution using a spectrophotometer or colorimeter.
- Investigate factors that impact on solution color intensity as measured by a spectrophotometer or colorimeter.
- Determine the relationships between color intensity, solution concentration, and wavelength.

PROCEDURES

In order to collect data quickly, different teams will study different identified salts found in the blood of the organisms (Table 1) and share results.

Table 1 Team Assignments: Salt Organism Species

Salt	Typical Reagent Formula
$Co(NO_3)_2$	$Co(NO_3)_2 \bullet 6H_2O$
$Cu(NO_3)_2$	$Cu(NO_3)_2 \bullet 2.5H_2O$
$Ni(NO_3)_2$	$Ni(NO_3)_2 \bullet 6H_2O$

PART I—PREPARATION OF SOLUTIONS OF DIFFERENT COLOR INTENSITY

Teams will prepare salt solutions of different color intensity using one of the identified salts found in the blood of the organisms (Table 1). When you are done, save your prepared solutions for studying the factors that impact on solution color intensity (Part II).

1. Record the formula of your assigned salt. Typical reagent formulas are provided in Table 1. Check the formula on the reagent bottle since hydrated salts often are available with different mole ratios of water to salt. Based on the formula, calculate the mass of salt needed to prepare 100 mL of 0.10-M solution. Do individual calculations and check your calculations with your teammates.

2. Measure and record the mass of salt to the nearest 0.01g. *Note:* It is difficult to measure a precalculated mass accurately. All that you need to do is to measure a known amount that is close to your calculated value. Then, calculate the *actual* concentration (M) of your sample based on the actual mass of salt that you use.

3. Transfer the salt sample into a 100-mL volumetric flask (see Appendix A). Add *a little* water and swirl the flask until the solid dissolves completely. Add more water until the bottom of the meniscus just touches the flask's calibration mark when your eye is level with it. Record the color of your prepared solution.

4. Prepare a minimum of five samples of lower color intensity by diluting your prepared 0.10-M solution. Obtain two burets to deliver different sample to distilled water ratios. Each buret should be secured in a proper buret stand when in use (see Appendix A). One buret will be used to deliver distilled water and the other to deliver your prepared 0.10-M solution. Prepare different sample to distilled water ratios (e.g., 10.00 mL to 0.00 mL; 7.00 mL to 3.00 mL; 5.00 mL to 5.00 mL; 4.00 mL to 6.00 mL; 3.00 mL to 7.00 mL). Record the ratio of sample to distilled water ratios below.

5. Calculate the concentration (M) of each diluted sample based on the ratio of sample to distilled water volume and enter the results in Table 2. Check your calculations with your teammates. Label the concentration (M) of each sample.

TEAM DATA

Table 2 Molarity of Samples of _____ of Different Color Intensity

Sample	Water (mL)	Sample (mL)	Solution M
1			
2			
3			
4			
5			

PART II— COLOR INTENSITY AND LIGHT INTERACTIONS

You are now to determine the factors that impact on solution color intensity as measured by a spectrophotometer or colorimeter. Each team is to conduct the investigation the salt solution assigned in Part I. Factors to investigate include the wavelength (λ) of light, solution concentration, and solution color.

1. In order to share data from the different teams, record in Table 3, at minimum, the amount of light absorbed by your salt solutions of different concentrations at wavelengths 450, 525, and 600 nm. (*Note:* If time allows, measure the light absorbed at one or more other wavelengths.)

TEAM DATA

Salt: _____   Solution color: _____

Table 3 Light Absorbance and Concentration at Different Wavelengths

Concentration (M)	Abs. λ 450 (blue light)	Abs. λ 525 (green light)	Abs. λ 600 (red light)	Abs. λ _____

2. A graph of absorbance versus concentration at a constant wavelength is called a calibration curve. Produce calibration curves at wavelengths of 450, 525, and 600 nm. Attach your graph(s). Compare the graphs. Identify the wavelength that gives the most quantifiable results.

3. The different teams should exchange data on the solution color and the absorbance at the tested wavelengths for 0.10 M solutions. Compile the different data using a computer or Table 4.

COMPILED TEAM DATA

Table 4 Light Absorbance of 0.10 M Salt Solutions

Salt	Solution Color[*] (V, B, G, Y, O, R)	Abs. λ 450 (blue light)	Abs. λ 525 (green light)	Abs. λ 600 (red light)	Best λ[#]
$Co(NO_3)_2$					
$Cu(NO_3)_2$					
$Ni(NO_3)_2$					

[*] V = violet, B = blue, G = green, Y = yellow, O = orange, R = red.
[#] Wavelength that gives the most quantifiable results.

Data Analysis and Implications (Part II)

1. Is there a pattern with regard to salt solution color and visible light absorption? For example, does the visible color of the solution relate to the wavelength color giving the highest or lowest absorbance values for the different solution concentrations? Support your decision using specific examples from the compiled team data in Table 4.

2. Does wavelength affect the amount of light absorbed by a solution? Refer to the team calibration curves produced at the different wavelengths (450 nm, 525 nm, and 600 nm). Describe any similarities or differences in results at the different wavelengths.

3. Use your team calibration curves to determine:

   a. Is there a mathematical relationship between absorbance and concentration of the solution? If so, indicate the relationship.

b. Suppose you had a solution of unknown concentration of the same salt. Describe how you would determine its concentration from its absorbance (for example, Abs. = 0.65 at the wavelength for the calibration curve giving the maximum absorbance values) and the mathematical relationship described under a. and/or your team calibration curve.

One of the central problems in chemistry is solution analysis. What is in a solution and how much of it is there? The wastewater and sewage treatment plant will provide you with a sample of water containing an unknown concentration of one of the colored salts found in the blood of the unique organisms living in the sewer. Your task is to determine the concentration (molarity) of the sample by measuring the color intensity with a spectrophotometer or colorimeter.

GUIDELINES

- Refer to the team-compiled data in Table 4 for selecting a wavelength for your analysis.
- If the unknown's absorbance reads outside your calibration curve values:
  - Do not extend the calibration curve line beyond known absorbance versus concentration values.
  - Dilute the unknown so its absorbance reads within the range of your curve OR
  - Do consider repeating the study at a different wavelength.
- Remember to correct for any dilution factor when calculating the true concentration.

PROCEDURE AND DATA

Experiment 3–3 Report

Provide a report summarizing your findings with regard to the factors that affect color intensity. Make sure to refer to specific data with regard to the color, concentration, and visible light interactions of the salt samples to support your conclusions.

1. Humans and other multicell organisms have blood that contains salts. For example, tunicates (a marine organism) have blood that contains approximately 0.1 M sulfuric acid, together with (depending on the species) approximately 0.1 M concentrations of vanadium or iron salts. Search the Web for information on the color of human blood and its salt composition. What gives human blood its color? What percentage of human blood contains salts? What are the primary salts in human blood?

2. A sample has an absorbance of 0.15 at λ 675. What is its concentration (mM) based solely on the calibration graph in Figure below?

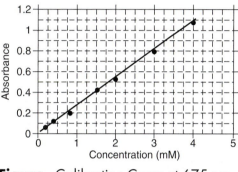

Figure Calibration Curve at 675 nm

Questions 3 and 4 refer to the absorption spectrum for a 0.40 M solution of metal ion M^+ and a calibration curve for the metal ion M^+.

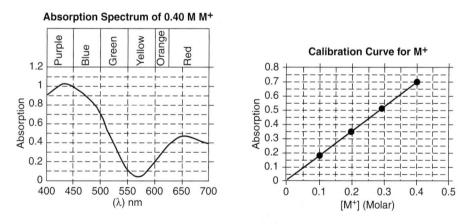

3. At what wavelength was the calibration curve obtained?

4. A sample of this solution of unknown concentration of M^+ has an absorbance reading that is "off scale" when read at the wavelength that was used to construct the above calibration curve. You dilute the sample by adding 5.0 mL of water to 1.0 mL of the solution of M^+. The absorbance of the diluted sample when read at the wavelength of the calibration graph = 0.35. What is the concentration (M) of the original undiluted sample?

Experiment 4–1

What Factors Affect the Solubility of Ions?

INTRODUCTION

A local company plans to produce test kits for identifying ions in aqueous solutions. They have hired research teams to study the solubility chemistry of ions. Teams are to collaborate and develop a plan for identifying metal ions in aqueous solution based on solubility data. The company also wishes to know what factors affect ion solubility. When you are done with your investigation, provide a report summarizing your findings.

OBJECTIVES

- Become familiar with the design of tests and the logic used to determine the identity of reacting ions forming precipitates.
- Collect and analyze data on the solubility behavior of metal ions in aqueous solution to determine whether metal ion solubility is predictable from metal ion characteristics.
- Design experiments to determine the identity of metal ions in a sample of water based on collected solubility data.

PROCEDURES

The research plan is to:

1. Become acquainted with the design of tests and the logic used for identifying ions in solution (Part I).
2. Different teams investigate and share data about the solubility of different ions to determine what factors affect ion solubility (Part II).
3. Develop and test a plan to identify metal ions in aqueous solution (Part III).

PART I—WHAT IS THE PRECIPITATE IDENTITY?

When the positive metal ion (cation) of a dissolved salt combines with the negative ion (anion) from a different dissolved salt, the recombined ions may either stay in solution or come out of solution in the form of a solid called a "precipitate." In this part of your inquiry, you will become familiar with the design of tests and the logic used to determine the identity of the reacting ions forming a precipitate. Ions that are present and do not precipitate but remain in solution are called spectator ions.

Part IA—Reaction of Calcium Chloride and Sodium Oxalate

You will analyze the reaction that occurs upon mixing solutions of calcium chloride and sodium oxalate: $CaCl_2 + Na_2C_2O_4 \rightarrow$ precipitate. Your goal is to determine the identity of the precipitate by conducting tests using other solution mixtures containing three of the four ions in the calcium chloride and sodium oxalate reaction and observing whether the precipitate forms or does not form. The resulting observations will allow you to identify the reactant ions forming the precipitate in the calcium chloride and sodium oxalate reaction.

> *Caution:* Do not dump any of the reagents down the sink.
> Discard the waste in an appropriate waste container.
> Do not allow the solutions to come in contact with your skin.

1. Obtain 5 mL of 0.10 M calcium chloride, $CaCl_2$, and 5 mL of 0.10 M sodium oxalate, $Na_2C_2O_4$. Record the appearance of each individual solution and the combined mixture in the table below. Label and save the mixture for reference.

Reagent 1 Ions	Reagent 2 Ions	New Ion Combos Possible After Mixing	Observations (precipitate?)
$CaCl_2$ $Ca^{2+}(aq) + Cl^-(aq)$	$Na_2C_2O_4$ $Na^+(aq) + C_2O_4^-(aq)$	$Ca^{2+} + C_2O_4^{2-}$ $Na^+ + Cl^-$	

2. Conduct tests of the reacting system to determine the identity of the ions forming the precipitate. Note that test solutions are the same concentration (0.10 M) as that in the reacting system under analysis. In addition, identical volumes (5 mL) are combined for the tests. Fill in the table below as you conduct the indicated tests.

Test	Reagent 1 Ions	Reagent 2 Ions	New Ion Combos Possible	Observations (precipitate?)
1.	0.10 M $CaCl_2$ $Ca^{2+} + (aq) + Cl^-(aq)$	0.10 M $NaNO_3$ $Na^+(aq) + NO_3^-(aq)$		
2.	0.10 M $Ca(NO_3)_2$ $Ca^{2+}(aq) + NO_3^-(aq)$	0.10 M $Na_2C_2O_4$ $Na^+(aq) + C_2O_4^{2-}(aq)$		
3.	0.10 M $NaCl$ $Na^+(aq) + Cl^-(aq)$	0.10 M $Na_2C_2O_4$ $Na^+(aq) + C_2O_4^{2-}(aq)$		

3. Compare your test results with the results from the reaction of calcium chloride and sodium oxalate. Identify and record below any ion combinations that stay in solution together and thus must be spectator ions in the reaction.

4. Based on the above tests, and ruling out ion combinations that are soluble and stay in solution, what is the likely identity of the precipitate in the reaction of calcium chloride and sodium oxalate?

5. Check the properties of your proposed precipitate in the Physical Constants of Inorganic Compounds section of the *CRC Handbook of Chemistry and Physics.* Do the listed properties match the observable properties of the precipitate? Congratulations. The precipitate you have identified forms from solutions in the bodies of people who suffer from kidney stones.

Part IB—Reaction of Calcium Chloride and Sodium Sulfate

Design tests and use your logic to confirm the identity of the precipitate that forms upon mixing solutions of calcium chloride and sodium sulfate: $CaCl_2 + Na_2SO_4 \rightarrow$ precipitate.

1. Obtain 5 mL of 0.10 M calcium chloride, $CaCl_2$, and 5 mL of 0.10 M sodium sulfate, Na_2SO_4. Record the appearance of each individual solution and the combined mixture. Label and save the mixture for reference.

2. Based on your logic and the information gained in Part IA about the solubility of some ion combination, what is the likely identity of the precipitate? Record your hypothesis.

Hypothesis

3. Design and carry out tests to determine and confirm the identity of the ions forming the precipitate. Create a table similar to the table in Part IA for recording your tests and observations. For example, can you design a test that omits sulfate ion in order to determine whether it is critical to precipitate formation?

4. Do your proposed products make sense? Geologists call the precipitate gypsum. Check the properties of your proposed products in the *CRC Handbook of Chemistry and Physics* or another chemistry reference text.

Data Analysis and Implications (Part I)

INFORMATION

- The formation of a precipitate in a chemical reaction is indicated by placing (*s*) next to the solid coming out of solution.
- The symbol (*aq*) is placed next to ions or salts remaining in solution.
- A net ionic equation omits spectator ions and shows only the reacting ions producing the precipitate.

1. For each of the reactions investigated in Parts IA and IB, write:

    a. A chemical equation that represents the reaction, includes both spectator and reactant species, and reflects your data.
    b. A complete ionic equation.
    c. A net ionic equation that shows only the reacting ions producing the precipitate.

    Part IA—Reaction of Calcium Chloride and Sodium Oxalate

    a.

    b.

    c.

    Part IB—Reaction of Calcium Chloride and Sodium Sulfate

    a.

    b.

    c.

2. Identify any ions present in both reacting systems that are spectator ions.

Why do some combinations of ions stay in solution, while others precipitate? Is the solubility of different ion combinations predictable? You will work in teams to collect and share data about the solubility of different ion combinations and to interpret results.

Table 1 and the information below indicate the different cations to be assigned and investigated by teams. Different team data will be compiled and the results used to explore possible links between structure and solubility.

Table 1 Team-Assigned Cations

I	Na^+	Ba^{2+}	Mg^{2+}	Co^{2+}	Ni^{2+}	Cu^{2+}	Cd^{2+}	Al^{3+}
II	K^+	Li^+	Ca^{2+}	Sr^{2+}	Cr^{3+}	Fe^{3+}	Zn^{2+}	Ag^+

INFORMATION

- All teams are to use 0.10 M nitrate salts of the assigned cations in order to ensure that solubility differences are the effect of the different cations.
- All teams are to use 0.10 M sodium salts of the same anions (NO_3^-, Cl^-, CrO_4^{2-}, I^-, $C_2O_4^{2-}$, CO_3^{2-}, SO_4^{2-}) in order to ensure that solubility differences are the effect of the different anions.

Solubility and Precipitation

For a salt to dissolve, ionic bonds must be broken and reformed, involving changes in energy. In the solid, the ions are fixed in a rigid lattice, while in solution the ions are mixed with water molecules and free to move about in solution. A water–ion attraction cloaks each ion on the surface of the solid with water molecules, and the ions are pulled into the water phase. At the same time the orientation of the water molecules about the ions (Figure 1) due to water–ion attraction reduces the water molecules' freedom of movement. When a salt precipitates, the process of dissolving is reversed.

1. Before starting the experiment, study Figure 1 and the information provided above about salt solubility and precipitation. Do you expect solubility to be linked to a particular ion characteristic (e.g., magnitude of charge, size, or something else)? Record your hypothesis.

Hypothesis

Figure 1 Orientation of H_2O Molecules about Ions

If your hypothesis is correct, which four of your eight assigned cations are *least* likely to form precipitates?

2. Table 2 provides a full-page grid sheet for conducting the experiment. Alternatively, your instructor may ask you to use cell well plates. If using the Table 2 grid, write the symbols of your team-assigned cations in column 1. Place an acetate sheet or a glass plate over the grid sheet.

3. Add one drop of your first assigned cation solution (0.10 M sodium nitrate, $NaNO_3$ or 0.10 M potassium nitrate, KNO_3) to each column of the first horizontal row. Add one drop of the appropriate cation solution to each column of the remaining horizontal rows.

> *Caution:* Do not allow your skin to be exposed to the solutions. Silver ion, Ag^+, will discolor your skin. Some ions are toxic.
> Do not discard any of the reagents down the sink.
> Do discard the waste in an appropriate waste container.

4. Add one drop of the first anion solution (0.10 M sodium nitrate, $NaNO_3$) to each row of the first vertical column. Add one drop of the appropriate anion solution to each row of the remaining vertical columns. Take care not to allow the dropper tip to contact the cation solution drop or you will contaminate the anion solution!

5. Record your team data in Table 3. Record a (P) if a precipitate formed and indicate the color (W = white, Y = yellow, O = orange, R = red, B = blue, G = green, Gr = gray, Bl = black). Record an (S) for soluble if there was no precipitate. When you have recorded your data, compare your results with any other team testing the same group of compounds. If necessary, repeat your tests. Do not discard your experiment grid sheet! Save it to refer to while conducting Part III. After completing Part III, rinse the acetate sheet, glass plate, or cell well plates, whichever you used, with water into a waste container as indicated by your instructor.

6. Does your team data on Table 3 validate or contradict your hypothesis regarding the four cations least likely to produce precipitates?

7. Share and collect the results of the different teams. Compile the different team data using a computer or Table 4 or as indicated by your instructor.

Table 2 Team Cation and Anion Sets (Part II)

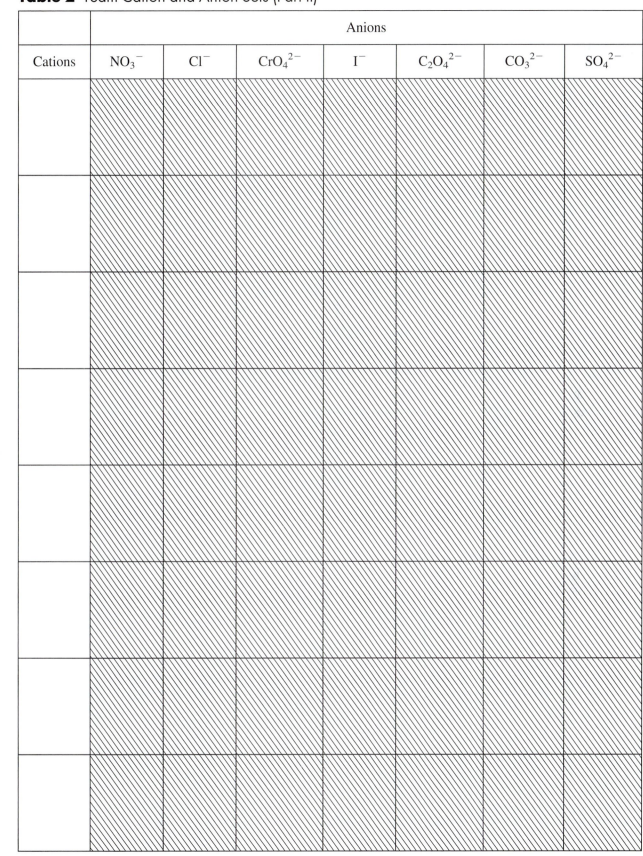

Cations	Anions						
	NO_3^-	Cl^-	CrO_4^{2-}	I^-	$C_2O_4^{2-}$	CO_3^{2-}	SO_4^{2-}

Table 3 Precipitation of Cations

Cations	Anions						
	NO_3^-	Cl^-	CrO_4^{2-}	I^-	$C_2O_4^{2-}$	CO_3^{2-}	SO_4^{2-}

Abbreviations used in table:
(S) = soluble, (P) = precipitate.
Color : W = white, Y= yellow, O = orange, R = red, B = blue, G = green, Gr = gray, Bl = black.

Table 4 Precipitation of Cations

Cations	Anions						
	NO_3^-	Cl^-	CrO_4^{2-}	I^-	$C_2O_4^{2-}$	CO_3^{2-}	SO_4^{2-}
Na^+							
Ba^{2+}							
Mg^{2+}							
Co^{2+}							
Ni^{2+}							
Cu^{2+}							
Cd^{2+}							
Al^{3+}							
K^+							
Li^+							
Ca^{2+}							
Sr^{2+}							
Cr^{3+}							
Fe^{3+}							
Zn^{2+}							
Ag^+							

Abbreviations used in table:
(S) = soluble, (P) = precipitate.
Color : W = white, Y= yellow, O = orange, R = red, B = blue, G = green, Gr = gray, Bl = black.

Data Analysis and Implications (Part II)

3. Identify the spectator ions (cation and anion) that are common to all your test solutions.

4. Write net ionic equations based on the compiled data for the combination of the cation reagent, $Ba(NO_3)_2$, with the different anion reagents.

5. Is it possible to predict whether a precipitate will likely be white or a color other than white based on the position of the cation's element in the periodic table? Refer to a periodic table and the compiled data of the different teams with regard to the precipitate color of the cations that reacted to support your conclusion.

Data Analysis and Implications (Part II continued)

6. What generalizations, if any, can be made about the solubility of ionic compounds with singly-charged alkali metal cations (Li^+, Na^+, and K^+)? What generalizations can be made about the solubility of ionic compounds with singly-charged anions (Cl^- and I^- and NO_3^-)? Refer to the compiled solubility data to support your conclusions.

7. What generalizations, if any, can be made about the solubility of ionic compounds with multiply-charged anions? Compare the solubility data of multiply-charged anions (CrO_4^{2-}, $C_2O_4^{2-}$, CO_3^{2-}, and SO_4^{2-}) with the singly-charged ions investigated in Question 6.

8. What generalizations, if any, can be made about the solubility of an ionic compound, if both the cation and anion are multiply-charged? Refer to examples from the compiled data to support your reasoning.

Data Analysis and Implications (Part II continued)

9. Figure 2 gives the ionic radii of some common metal ions.

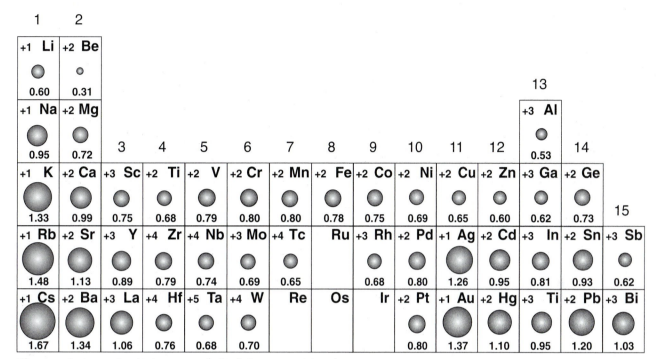

Figure 2 Ionic Radii (picometers) of Common Metal Ions

What generalizations, if any, can be made about solubility and metal ion size (radius)? Compare the number of precipitates produced with the different anion combinations for the tested alkaline earth cations (Mg^{2+}, Ca^{2+}, Sr^{2+}, Ba^{2+}) of group 2 and the cations (Zn^{2+} and Cd^{2+}) of group 12. Organize and refer to the data to answer this question.

Your task is to use the results and compiled solubility data of the different teams from Part II to design an experiment to identify ions in solution. You will be assigned to one of the problems (A or B) below.

Part IIIA—What Is the Identity of the Metal Ion in the Well Water?

Each team will be given a sample of different well water. Each sample contains one of the following ions: Ba^{2+}, K^+, Al^{3+}, Cu^{2+}, Ca^{2+}, or Sr^{2+}. You are to determine experimentally which ion is present. Record your procedure and results below.

Part IIIB—What Is the Identity of the Anion in the Solution?

Your challenge is to design a reaction procedure to determine which of four anions (SO_4^{2-}, $C_2O_4^{2-}$, CrO_4^{2-}, or CO_3^{2-}) is present in an aqueous solution. Record your procedure and results below.

TEAM PROCEDURES AND RESULTS

Experiment 4–1 Report

Provide a report summarizing your solubility findings. Provide suggestions for the test kits for identifying metal ions in aqueous solution that the company plans to produce. Provide examples from the experimental data to support your solubility conclusions and test kit suggestions.

1. If a 0.10 M solution of silver fluoride, AgF, is mixed with a 0.10 M solution of barium acetate, $Ba(C_2H_3O_2)_2$, a precipitate (ppt.) forms: $AgF + Ba(C_2H_3O_2)_2 \rightarrow$ ppt.

   In order to identify the precipitating ions, you conduct some experimental tests.

   a. What is the purpose of this test (the solutions are 0.1 M): $AgF + NaC_2H_3O_2 \rightarrow$?

   b. What do the test results below tell you about the $AgF + Ba(C_2H_3O_2)_2$ reaction?

      $AgF + Ba(NO_3)_2 \rightarrow$ precipitate (Properties are identical to those of the ppt. formed in the $AgF + Ba(C_2H_3O_2)_2$ reaction.)

   c. Which 0.10 M test reagent combination below did you use to *directly* test the hypothesis that silver ions are critical to precipitate formation?

      $AgF + Ba(NO_3)_2$ $NaF + Ba(C_2H_3O_2)_2$ $AgF + KC_2H_3O_2$

2. Your team investigates the reaction: $BaCl_2(aq) + AgF(aq) \rightarrow$ white precipitate. A teammate observes: $FeCl_3(aq) + AgF(aq) \rightarrow$ light green precipitate. This observation tells you nothing about the identity of the white precipitate. Why?

3. Choose the salt compound within each pair that is a color other than white, and indicate the reasoning for your decision.

   a. Nickel sulfate or lead sulfate
   b. Cobalt oxalate or strontium oxalate

4. Check the "solubility rules" in a chemistry textbook. How do the rules compare to the generalizations about solubility made from the compiled team solubility data?

5. A solution is either 0.10 M $Li(NO_3)$ or 0.10 M $Ba(NO_3)_2$, or 0.10 M $Cu(NO_3)_2$. When you add 1 mL of 0.10 M K_2CO_3 a white precipitate forms. What is the identity of the solution?

6. Assume you had a solution that contained a mixture of silver ions and strontium ions. Come up with three different reaction plans that would allow you to precipitate either the silver ion or strontium ion, but the other ion would remain in solution.

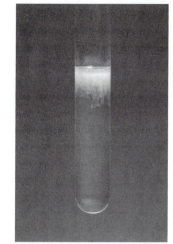

Experiment 4–2

Can Toxic Ions Be Removed from Water by Precipitation?

A local company has successfully produced and marketed test kits for identifying ions present in aqueous solutions. The company is interested in expanding its services to include the removal of toxic ions from aqueous solution. The company wishes to know whether ions can be completely removed from water by precipitation. Is some minimum concentration of ions required for precipitation? Does the presence of solvents other than water effect the solubility and precipitation of ions?

OBJECTIVES

- Collect and analyze data on the precipitation of ions in aqueous solution to determine whether there are links between ion concentration and precipitation.
- Collect and analyze data on the precipitation of ions in aqueous solution to determine whether ions can be completely removed from water by precipitation.
- Compare the effect of water and solvents other than water on salt precipitation.

PROCEDURES

Your team will collect data about the effect of solvents (Part I) and concentration (Part II) on salt solubility. Each team will investigate a different group of instructor assigned salts (Table 1) and share results in order to collect data quickly. Based on results you will design a procedure for removing toxic ions from solution (Part III).

Table 1 Team Salts for Solvent Studies

Group	Salt #1	Salt #2
1	$Zn(NO_3)_2$	NaOH
2	$Ca(NO_3)_2$	$Na_2C_2O_4$
3	$Cd(NO_3)_2$	Na_2CO_3

PART I—EFFECT OF SOLVENTS ON SALT SOLUBILITY

Polar molecules such as water have a separation of charge with a positive and negative pole. Figure 1 depicts the polarity of water molecules where the H end of the molecule is positively charged and the O end is negatively charged. How critical is polarity to salt solubility? In this investigation, you will compare the solubility and precipitation of salts in solvents that differ in polarity.

Salt Solubility and Water

Figure 1 Orientation of Polar Water Molecules about Ions

When a salt dissolves, the partial positive and negative charges of water molecules provide a substitute for the charges (ions) in the solid salt. The negative poles of some water molecules attract the positive ions (cations) while the positive poles of some water molecules attract the negative ions (anions) in the solid salt. This water–ion attraction surrounds each ion on the surface of the crystal with water molecules, and the ions are pulled into the water phase. Figure 1 depicts the orientation of the water molecules about a cation and anion.

INFORMATION

Water (H_2O) is polar; acetone (CH_3COCH_3) is moderately polar; hexane (C_6H_{14}) is nonpolar.

1. You will test and compare the solubility and precipitation of salts in water, acetone, and hexane. Record your hypothesis regarding the effect of solvent polarity on solubility and precipitation of salts. If your hypothesis is correct, what will you observe? For example, will precipitation occur in all solvents?

    Hypothesis

2. To compare the solubility of your assigned salts in water, acetone, and hexane, add a few crystals of salt #1 to each of three separate dry test tubes labeled water, acetone, and hexane. Add 2.0 mL (40 drops) of water, acetone, or hexane to the crystals in the appropriate tube. Mix vigorously. Record your observations in column 2 of Table 2. Are the crystals "insoluble," "slightly soluble," or "soluble" in each testing solvent? Save the labeled samples for later use. Repeat the tests with salt #2 and record your observations in column 3 of Table 2. Save the labeled samples for later use.

Caution: Do not inhale acetone or hexane. Keep the samples closed or use a hood.
Do not dispose of acetone or hexane down the sink.
Discard the acetone or hexane in an appropriate (organic) waste container.

TEAM DATA

Salt 1: _____   Salt 2: _____

Table 2 Salt Solubility and Precipitation in Different Solvents

Solvent	Solubility Salt #1 (soluble, insoluble, slightly soluble)	Solubility Salt #2 (soluble, insoluble, slightly soluble)	Precipitation (a lot, slight, none, unsure)
H_2O			
CH_3COCH_3			
C_6H_{14}			

3. Separate any undissolved solid from the solvent solution (containing dissolved crystal). For example, pipet or decant off the supernatant liquid (the clear solution above any undissolved solid) into **clean and dry** test tubes or containers (labeled to identify the contents).

4. Combine the clear liquids of salts #1 and #2 from the same solvent mixture (e.g., salt #1 mixed with water and salt #2 mixed with water). Does precipitation occur? How much? Record "a lot," "slight," "none," or "unsure" in Table 2. When you are finished, compare your results with any team testing the same salts. Repeat tests if there are disagreements in results.

5. Do your team results show any pattern of results between precipitation (step 4 above) and solubility (step 2) of the salts in a given solvent?

6. What will happen to salt solubility if you mix the different solvents? For example, what will happen to salt solubility if you add acetone or hexane to an aqueous salt solution? Place about 3 mL (60 drops) of a saturated solution of common table salt, NaCl, into two separate small test tubes. To the first tube add about 10 drops of acetone. To the second tube add about 10 drops of hexane. Record your observations.

7. Enter your team's data into the class database. Use a computer or Table 3. Obtain a copy of the compiled data of the different teams.

Table 3 Solubility and Precipitation in Different Solvents

| Group | Solubility* | | | | | | Precipitation** | | |
| | Salt 1 | | | Salt 2 | | | Salt 1 + 2 | | |
	Water	Acetone	Hexane	Water	Acetone	Hexane	Water	Acetone	Hexane
1									
2									
3									

*I = Insoluble, S = soluble, Sl = slightly soluble.

**L = a lot, Sm = slight amount of precipitate, No = none, U = unsure.

Data Analysis and Implications (Part I)

1. Are there any visible patterns in the compiled data of the different teams regarding solvent polarity and solubility of salts? Record any generalizations that can be made about solvent polarity and salt solubility.

2. Are there any visible patterns in the compiled team data between amount of precipitate (none, slight, a lot) and solvents (water, acetone, hexane) of different polarity? Organize the data in the form of a table or graph to answer this question and record your conclusion.

3. Are there any visible patterns in the compiled data of the different teams linking solvent polarity, salt solubility, and precipitation? For example, if salts are highly soluble in a particular solvent, does the combined salt mixture produce more or less precipitate? Refer to specific data examples to support your conclusion.

Is some minimum concentration of ions required for precipitation? Can ions be completely removed from water by precipitation? Your task is to investigate precipitation at different ion concentrations and collect data to answer these questions. In order to quickly collect data, teams are to use aqueous solutions of the same salts used in Part I (Table 1) and share results.

1. Transfer 10 drops of a 0.10 M solution of assigned salt #1 to a small test tube. Add 10 drops of a 0.10 M solution of assigned salt #2. **Shake well.** Record the amount of precipitate as ''a lot,'' ''slight,'' ''none,'' or ''unsure.'' Identify the reacting ions and products. Label and save the precipitated mixture.

2. Repeat the reaction using 10 drops of both reactant salts at $\approx equal$ concentrations above and below 0.10 M. In Table 4, record the team consensus about the amount of precipitate as ''a lot,'' ''slight,'' ''none,'' or ''unsure.'' Label and save any precipitated mixture.

INFORMATION

- If < 0.10 M reagent is not available, dilute the 0.10 M sample (for example, 2 drops of 0.10 M reagent + 18 drops of distilled water = 0.010 M).
- If > 0.10 M reagent is not available, add a few crystals of the solid to its 0.10 M solution.

TEAM DATA

Salt 1: _____   Salt 2: _____

Table 4 Precipitate Yield* versus Concentration

< 0.10 M	0.10 M	> 0.10 M

*Record ''a lot,'' ''slight,'' ''none,'' or ''unsure.''

4. Have all the ions precipitated at the different concentrations? Filter the precipitated mixtures at the different concentrations and collect the filtrate (the *clear* liquid passing through the filter). Divide the filtrate into two portions.

5. Conduct tests to determine whether there are metal ions that failed to precipitate present in the filtrate. To one filtrate portion, add a few crystals of salt #2 and stir. To the other portion, add 1–5 drops of 0.10 M Na_2S. (*Note*: Most transition and heavy metal ions form insoluble sulfide salts.) Does a precipitate form? Record your observations in Table 5.

TEAM DATA

Salt 1: _____   Salt 2: _____

Table 5 Filtrate Observations

Concentration	Filtrate + Crystals	Filtrate + 0.1 M Na_2S
< 0.10 M		
0.10 M		
> 0.10 M		

6. Compare your results with those of any other team using the same salt combinations. In the event of differences, repeat tests as necessary.

7. Share your data and collect the compiled data of the different teams. Use Table 6 or a computer to compile the different team data as indicated by your instructor.

Table 6 Concentration and Precipitation[*]

Group	Concentration		Filtrate		
	< 0.1 M	0.1 M	> 0.1 M	+ Crystals	+ 0.1 M Na$_2$S
1					
2					
3					

[*]L = a lot, Sm = slight amount of precipitate, No = none, U = unsure.

Optional Extensions

1. What will you observe if the salts are not equal in concentration? For example, what will happen to the amount of precipitate formed, if salt 2 is at a concentration greater than 1.0 M while salt #1 remains at 0.10 M concentration?

2. Does the order of addition of salt reactants effect the results? For example, add

   5 drops of 0.1 M Zn(NO$_3$)$_2$ to 5 drops of 1 M NaOH $\rightarrow$?
   5 drops of 0.1 M NaOH to 5 drops of 0.1 M Zn(NO$_3$)$_2$ $\rightarrow$?

Data Analysis and Implications (Part II)

4. What generalizations, if any, can be made about salt concentration and precipitation? Refer to the compiled precipitation and concentration data to support your conclusions.

5. Does precipitation of the different salt combinations occur within the same concentration or different concentration values? Refer to examples in the compiled team data to support your answer.

6. You need to remove an "undesirable ion" from solution. Is it possible to completely remove the ion from solution by precipitation and filtration? Refer to examples in the compiled team data to support your answer.

The water effluent from a plant contains a mixture of Cu^{2+} and Na^+ ions. You are to design a method for maximizing removal of Cu^{2+} ions from the mixture. You may use any of the reagents (precipitating agents or solvents) tested in this experiment and/or filtration. Make sure to record your procedure and observations below.

INFORMATION

- You might find it useful to test the solubility of Cu^{2+} ions with the different precipitating agents used in this experiment.
- Solubility data regarding Cu^{2+} ions in combination with the different precipitating agents used in this experiment can be found in reference texts such as the *CRC Handbook*.
- If desired, the sample's absorbance can be read at λ_{max} (λ 650 nm) using a spectrophotometer. A reading of zero indicates removal of the Cu^{2+} ions.

PROCEDURE AND OBSERVATIONS

Experiment 4–2 Report

Submit a statement to the company indicating whether toxic ions can be completely removed from water by precipitation. Support your statement with experimental data examples and include answers to the following questions:

- Is some minimum concentration of ions required for precipitation?
- Is the concentration of the precipitating agent important?
- Does the presence of solvents other than water effect the precipitation of ions?

1. A gram of finely divided salt dissolves much faster than a single crystal of salt with a mass of one gram. Explain why.

2. Solutions containing ions are able to conduct an electric current, whereas solutions containing molecules cannot conduct current because there are no charge carriers present. Offer an explanation for the following observations:

   a. A mixture of solid NaCl + water is an excellent conductor of electricity, whereas a mixture of solid NaCl + hexane does not conduct electricity.
   b. An aqueous solution of saturated $HgCl_2$ is a much poorer conductor of electricity than an aqueous solution of saturated $CaCl_2$.

3. Do precipitation reactions go to completion or are they equilibrium systems? If the reactions go to completion, the reactants (reacting ions) completely form products (precipitate). If precipitation is an equilibrium system, both reactants and products are always present in the mixture. What conclusion do the data from this experiment support?

4. When 10 mL of 0.1 M indium (III) sulfate is added to 10 mL of 0.1 M titanium (III) chloride, a green precipitate of titanium (III) sulfate is formed:

   $$TiCl_3\,(aq) + In_2(SO_4)_3\,(aq) \rightarrow \text{green precipitate identified as}$$
   $$\text{titanium sulfate}$$

   You repeat the above reaction, but now add 10 mL of 0.001 M indium sulfate solution to 10 mL of 0.001 M titanium (III) chloride. No precipitate forms. Why?

5. The concentration of NaCl in Utah's Great Salt Lake is $\approx$ 6 M. If a drum containing acetone accidentally spills into the lake, is it likely that some NaCl will precipitate? *Information:* The solubility of NaCl = 35.7 g/100 mL (0°C) or 39.1 g/100 mL (100°C).

6. If $Na_2CrO_4\,(s)$ is gradually added to a solution containing 0.010 M Pb^{2+} and 0.010 M Ba^{2+}, $PbCrO_4\,(s)$ will precipitate first (rather than both $PbCrO_4$ and $BaCrO_4$ precipitating simultaneously). Offer an explanation.

Experiment 5–1

How Do Gas Volumes Change as We Ascend and Descend?

INTRODUCTION

Scuba divers may experience severe physical pain as they descend into and ascend from deep water due to pressure changes. For example, pressure changes can cause ear pain due to changes in the volume of air trapped in the diver's middle ear.

Temperature changes can also alter the volume of gases in hot-air balloons. For example, hot-air balloons are equipped with a heat source to change the volume of gas trapped in the balloon. Hot-air balloons ascend only if the density of the air in the balloon is less than the density of the surrounding air. If the density of the air in the balloon is greater than the density of the surrounding air, the balloon descends.

In this experiment, you will examine the effect of pressure and temperature on gas volume. Your goal is to determine what exactly are the effects of pressure changes on gas volumes such as the volume of air trapped in the middle ear of a scuba diver. How do temperature changes affect air density and the ascent or descent of hot-air balloons?

OBJECTIVES

- Determine how pressure changes as you alter the volume of a known amount of gas at constant temperature.
- Determine how volume changes as you alter the temperature of a fixed amount of gas at constant pressure.
- Represent the relationship of pressure–volume and temperature–volume results graphically and mathematically.

PROCEDURES

You will study the effect of pressure (Part I) and temperature (Part II) on gas volume. Your instructor will assign different teams to different gas volume studies. When you are done, results will be shared.

PART I—PRESSURE AND VOLUME STUDIES

To study the effect of volume changes on the pressure of a known amount of gas, you will be provided with a syringe containing trapped air that is attached to a gas pressure sensor. All teams will start with a volume of 5.0 mL of gas and record its pressure. The volume of gas

Table 1 Team Volume (mL) Changes

Team	mL of Gas Each Time
1	1.0
2	2.0
3	3.0

will be increased in increments and the pressure recorded for at least 5 or 6 readings or until the syringe reads 20 mL. Your instructor will assign teams to specific volume increases (Table 1). When you are done, data from the different teams will be shared.

1. If you increase the volume of a known amount of gas at constant temperature, what do you expect will happen to its pressure? Record your hypothesis.

   Hypothesis

2. Prepare the pressure sensor and an air sample for data collection:
   a. Plug the pressure sensor into Channel 1 of the computer interface.
   b. With the 20 mL syringe disconnected from the pressure sensor, move the piston of the syringe until the front edge of the inside black ring (Figure 1) is at the 5.0 mL mark.

Figure 1 Gas Syringe

   c. Attach the 20 mL syringe to the valve of the pressure sensor. (Some of the gas pressure sensors have a white stem protruding from the end of the sensor box—attach the syringe directly to the white stem with a gentle half-turn.)

3. Connect the computer to the LabPro interface using the USB cable and then connect the interface to the electrical outlet.

4. Turn on the computer and run the LoggerPro program.

5. Set up the computer and interface for the pressure sensor.

   a. Select EXPERIMENT from the menu bar.
   b. Select DATA COLLECTION from the EXPERIMENT menu.
   c. Select EVENTS WITH ENTRY from the MODE dialog box.
   d. Type "Volume" in the Column Name dialog box.
   e. Type "Vol" in the Short Name dialog box.
   f. Type "mL" in the Units dialog box.
   g. Press the Done button.

6. Click COLLECT to begin data collection.

7. Collect the pressure vs. volume data. One person in your team should take care of the gas syringe and another should operate the computer. Make sure that the piston of the front edge of the inside black ring (see Figure 1) is at the 5.0 mL line on the syringe. Hold the piston firmly in this position until the pressure value stabilizes. When the pressure reading has stabilized, click KEEP. Type "5.0" in the edit box. Press the ENTER key to keep this data pair. *Note:* You can choose to redo a point by pressing the ESC key (after clicking KEEP, but before entering a value).

8. Continue the above procedure for the assigned volume increase (6.0, 7.0, or 8.0 mL) and continue taking at least 5 or 6 pressure readings or until the volume reaches 20.0 mL.

9. Click STOP when you have finished collecting the data, which will appear in a table on the left-hand side of the screen. On the right side, you will see the data in graphical form.

10. Record in Table 2 the values given on the computer screen for pressure and volume. For each pressure–volume pair, calculate *P/V* and *PV*.

Table 2 Volume versus Pressure

Volume (mL)	Pressure (atm)	*P/V*	*PV*

11. Based on your data and calculations in this part, what is the relationship between the pressure and the volume of a gas sample? Do your team results contradict or support your hypothesis regarding the effects of increasing gas volume on pressure?

12. When you are finished, collect the volume and pressure data of the different teams using a computer or Table 3 provided.

Table 3 Volume versus Pressure

Volume	Team #	Pressure	P/V	PV

Date: _____ Name: _____ Class: _____

Data Analysis and Implications (Part I)

1. Which variables (volume, pressure, temperature) were held constant during this part of the experiment?

2. Prepare a graph of pressure vs. volume using the compiled team data. (Attach your graph). Based on the pressure vs. volume graph of the compiled team data:

    a. Describe the relationship between pressure and volume.

    b. Predict what would happen to the volume of a sample of gas if the gas pressure were to drop to one-third of the original value.

    c. Predict what would happen to the gas pressure of a known amount of gas if its volume were to decrease by one-third of its original value.

3. Prepare a graph of pressure vs. volume using the compiled team data, and draw the best straight line. (Attach your graph.)

   a. What type of relationship does it represent? Explain.

   b. Do the results offer a useful relationship and possible explanation for the problem under investigation: How does the volume of air trapped in the middle ear of divers change as they descend into and ascend from deep water due to pressure changes? Why? Explain.

You will use a temperature sensor to study the effect of temperature on the volume of a known amount of gas. Each team will study the increments in volume (for a given amount of gas) as the temperature increases, as assigned in Table 4.

Table 4 Team Volume Studies (mL)

Team	Volume Changes (mL)
1	1.0
2	1.5
3	2.0

1. If you increase the temperature of a sample of gas, what will happen to its volume? Assume that the gas sample is a constant mass of gas at constant pressure. Record your hypothesis.

   Hypothesis

2. Align the handle of the valve attached to the end of the syringe as shown in Figure 2. Fill the 20 mL syringe with air to the 10.0 mL mark (take the lower ring of the plunger as a reference) and close the valve, rotating it by 90 degrees as shown in Figure 3.

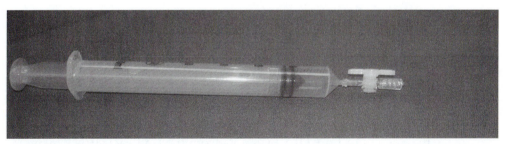

Figure 2 Syringe with Alignment of Attached Valve

Figure 3 Syringe with Closed Valve

3. Connect the temperature sensor to Channel 1 of the LabPro interface and then connect the computer to the interface using the USB cable as shown in Figure 4.

Figure 4 Computer and LabPro Interface

4. Connect the interface to the electrical outlet.

5. Turn on the computer and run the LoggerPro program.

6. Set up the computer and interface for the temperature sensor:
   a. Select EXPERIMENT from the menu bar.
   b. Select DATA COLLECTION from the EXPERIMENT menu.
   c. Select EVENTS WITH ENTRY from the MODE dialog box.
   d. Type "Volume" in the Column Name dialog box.
   e. Type "Vol" in the Short Name dialog box.
   f. Type "mL" in the Units dialog box.
   g. Press the Done button.

7. Prepare a water bath in a 1000 mL beaker (more than three-fourths full) and place it on a hot plate as shown in Figure 5. Insert the temperature sensor and the syringe into the water bath. Stir the water with a magnetic stirrer in order to have a uniform temperature. Wait for about five to eight minutes for the system to reach thermal equilibrium. Now you are ready to register temperature and volume data.

Figure 5 Water Bath on Hot Plate

8. Click COLLECT to begin data collection.

9. When the temperature reading has stabilized, click KEEP. Type "10.0" in the edit box. Press OK to keep this data pair. (At 10.0 mL, the volume in the syringe has been calibrated to 1 atm.)

10. Now turn on the hot plate and heat the water bath.

Caution: Do not touch the hot plate. Also do not allow the probe to touch the hot plate.
Be careful not to touch any hot surface or hot water with your bare hands. If it happens, immediately soak the area with ice and report it to your instructor.

11. Read the new volume of air in the syringe for the assigned increment (Table 4). Press KEEP and annotate in the edit box. Press OK to keep this data pair, (i.e., the volume increment and the corresponding temperature).

12. Continue with this procedure as the temperature of the bath increases to near boiling and you have at least 4 or 5 data pairs.

13. Press STOP when you have finished collecting data.

14. When you finish the temperature study, turn off the hot plate. Record in Table 5 the values given on the computer screen for volume and temperature. For each volume–temperature pair, calculate V/T and VT to complete the table.
    Note: Record the temperature in kelvins. Add 273° to the Celsius temperature to obtain the temperature value in kelvins.

TEAM DATA

Table 5 Volume versus Temperature

Volume (mL)	Temp. (°C)	Temp. (K)	V/T	VT

15. Based on your data and calculations in this part, what is the relationship between the volume and temperature of a gas sample? Do your team results contradict or support your hypothesis regarding the effects of increasing temperature on gas volume?

16. Collect the temperature and gas volume data of the different teams using a computer or Table 6.

COMPILED TEAM DATA

Table 6 Volume versus Temperature

Team #	Volume (mL)	Temperature (°C)	Temperature (K)	V/T	VT

Data Analysis and Implications (Part II)

4. Which variables (volume, pressure, temperature) were held constant during this part of the experiment? Why?

5. Prepare a graph of volume vs. temperature using the compiled team data. Use temperature values expressed in the Kelvin scale and answer the following questions:

   a. Describe any observed pattern of results between temperature changes and volume changes.

   b. Predict what would happen to the volume of a sample of gas if the temperature were to drop to one-half of the original value.

   c. Predict what would happen to the temperature of a known amount of gas if its volume were to decrease by one-third of its original value.

6. Do the results offer a useful relationship and possible explanation for the problem under investigation: How do temperature changes affect the volume of air contained in a balloon and the ascent or descent of hot-air balloons? Why? Explain.

Experiment 5–1 Report

Provide a brief report summarizing your results regarding how the volume of a gas is affected by its pressure and temperature. In particular, what are the effects of pressure changes on the volume of air trapped in the middle ear of scuba divers as they descend into and ascend from deep water? How exactly do temperature changes affect gas volume and the ascent or descent of hot-air balloons? Make sure to refer to experimental observations, data, and graphs to support your conclusions.

1. Charles' law (V/T = constant) describes the relationship between volume and temperature and Boyle's law (PV= constant) describes the relationship between volume and pressure. Search the Web and/or chemical literature for information on their research. Determine the mathematical expression that combines both Charles' and Boyle's laws.

2. A sample of carbon dioxide has a volume of 25.0 L at 5.80×10^2 mmHg. What pressure would be needed to increase the volume to 55.5 L? Assume constant temperature.

3. A sample of hydrogen occupies a volume of 950 mL at $-15.3°C$. If the temperature increased to 30.0°C, what would be the new volume?

4. A helium balloon is partially filled with 28.7 L of gas at 25.0°C and a pressure of 1.00 atm. If the balloon is released and it floats up into the atmosphere to a height where the pressure is 2.8×10^{-3} atm and the temperature is $-24.0°C$, calculate the new volume of the balloon.

5. The combustion of methane can be described by the equation

$$CH_2\,(g) + 2\,O_2\,(g) \rightarrow CO_2\,(g) + 2\,H_2O\,(g)$$

If 22.0 mL of $CO_2\,(g)$ were collected, what volume of $H_2O\,(g)$ also could be collected under the same conditions?

6. Apply the gas laws to determine how many grams of sodium azide (NaN_3) are needed to fill the air bag of your vehicle to a volume of 79.0 L at 35.0°C and 1.00 atm. The reaction occurring in the air bag can be represented by the equation, $2NaN_3\,(s) \rightarrow 2\,Na\,(s) + 3\,N_2\,(g)$. The gas constant, R, is equal to 0.0821 L atm mol^{-1}K^{-1}.

7. A hot-air balloon has a volume of 4.00×10^5 L at 30°C and a pressure of 748 mmHg. If the average molar mass of air is 29.0 g/mol, what is the mass of air in the balloon?

8. What is the volume of a scuba tank if it takes 2000 L of air collected at 1.00 atm to fill the tank to a pressure of 150 atm? T is constant.

Which Salts Make Good Cold Packs and Hot Packs?

INTRODUCTION

You have been hired as a consultant to a company that manufactures cold and hot packs, which are often used to treat an injury such as to a muscle or shoulder. The chemical reaction in the cold pack absorbs heat, while the chemical reaction in the hot pack releases heat.

The company recently received bequests of salts (chlorides, nitrates, and sulfates). Your job is to identify salts for use as reactants and provide guidelines regarding the mass of solid salt to be used in cold and hot packs. At the end of the investigation, you are to submit a report of your findings.

OBJECTIVES

- Identify salts that can serve as reactants in chemical cold and hot packs.
- Determine the heats of reactions for the different salts.
- Recommend the best salts for use in cold and hot packs based on the heat of reaction data.
- Determine any patterns between the heat behavior of different salts and salt mass.
- Provide recommendations regarding the mass of solid salt for use in cold and hot packs.

PROCEDURES

You and your teammates are to investigate the merits of different salts for potential use in cold or hot packs. Your supervisor will assign you to one of two research teams investigating different salts.

Team 1: Ammonium chloride (NH_4Cl), ammonium nitrate (NH_4NO_3), barium chloride ($BaCl_2$), magnesium sulfate ($MgSO_4$), potassium nitrate (KNO_3), sodium chloride ($NaCl$)

Team 2: Calcium chloride ($CaCl_2$), magnesium chloride ($MgCl_2$), magnesium nitrate ($Mg(NO_3)_2$), potassium chloride (KCl), sodium nitrate ($NaNO_3$)

Your preliminary studies are qualitative tests to determine whether heat is liberated or absorbed when a salt is dissolved in water.

1. Place about 10 mL of distilled water into a test tube and use a spatula to transfer about $1 \, cm^3$ or about 1 g of the assigned team salt to the test tube. Mix the contents with a stirring rod and touch the test tube in the palm of your hand to see whether it becomes warmer, cooler, or neither. Repeat this procedure for each assigned salt. Refer to the information below and enter your observations in Table 1 for each assigned salt.

INFORMATION

- For temperature change, enter increase, decrease, or neither.
- For type of reaction, enter endothermic if the mixture becomes cooler, exothermic if the mixture becomes warmer, or unsure.
- For the sign of q, enter $+$ if heat is absorbed by the mixture, $-$ if heat is liberated by the mixture, and $\pm$ if you detect no heat change.

TEAM DATA

Table 1 Qualitative Observations of Temperature Change

Salt	Temperature Change	Type of Reaction	Sign of q

2. Compare your results with any other team investigating the same salts. Repeat any team tests if there are discrepancies in the results. When you are done, collect the results of all teams using a computer or enter the results of the different teams in Table 2.

3. On the basis of the compiled results, select and record the three best potential salt candidates for hot packs and the three best potential salt candidates for cold packs.

Table 2 Qualitative Observations of Salt Temperature Changes

Salt	Temperature Change	Type of Reaction	Sign of q
NH_4Cl			
NH_4NO_3			
$BaCl_2$			
$Ca(NO_3)_2$			
$MgSO_4$			
NaCl			
$CaCl_2$			
$MgCl_2$			
$Mg(NO_3)_2$			
KCl			
$NaNO_3$			

The three best salts for hot packs are:

The three best salts for cold packs are:

Data Analysis and Implications (Part I)

For purposes of data analysis use the compiled class data (Table 2) and the information below.

INFORMATION

- When a salt dissolves in water, two opposite heat processes are involved. It requires heat energy to separate the associated ions in the solid salt, and heat energy is released when the water molecules surround (solvate) the ions. Whether heat is liberated or absorbed overall depends upon the sum of the energies associated with these two processes.
- The reaction of dissolving a salt in water (dissolution reaction) is represented below. MA represents the salt, where M is any metal ion ($+$) and A is the non-metal ion ($-$):

$$MA(s) \xrightarrow[\mathrm{H_2O}\,(l)]{} M^+(aq) + A^-(aq)$$

1. Complete dissolution reactions for all tested salts are:

   NH_4Cl (s) $\rightarrow$

   NH_4NO_3 (s) $\rightarrow$

   $BaCl_2$ (s) $\rightarrow$

   $MgSO_4$ (s) $\rightarrow$

   $NaCl$ (s) $\rightarrow$

   $CaCl_2$ (s) $\rightarrow$

   $MgCl_2$ (s) $\rightarrow$

   $Mg(NO_3)_2$ (s) $\rightarrow$

   KCl (s) $\rightarrow$

   KNO_3 (s) $\rightarrow$

   $NaNO_3$ (s) $\rightarrow$

Data Analysis and Implications (Part I continued)

2. This question deals with salts that are cold-pack candidates.

   a. List all salts yielding endothermic reactions that are candidates for cold packs.

   b. Based on the data, what process involves more energy in the endo-thermic reactions? Specifically, is more energy involved in separating the ions or solvating the ions? Explain.

3. This question deals with salts that are candidates for hot packs.

   a. List all salts yielding exothermic reactions that are candidates for hot packs.

   b. Based on the data, what process involves more energy in the exothermic reactions? Specifically, is more energy involved in separating the ions or solvating the ions? Explain.

Based on the observations in Part I, teams along with the instructor are to discuss and choose the three best candidates for cold packs and the three best candidates for hot packs for further quantitative investigations. Each team will then conduct two quantitative heat investigations using one cold and one hot pack candidate. For comparison purposes, each salt should be tested by a minimum of two teams.

THREE BEST SALTS FOR COLD PACKS

THREE BEST SALTS FOR HOT PACKS

ASSIGNED TEAM SALTS

Cold Pack Salt _____   Hot Pack Salt _____

1. For heat measurements, obtain two (rather than one) concentric Styrofoam cups to minimize heat exchange with the surroundings. Use two concentric Styrofoam cups with a thermometer/temperature probe and a glass/magnetic stirrer inserted into the cup through a cardboard or plastic lid. (See Figure 1.)

2. Using a graduated cylinder, transfer 100 mL of distilled water into the calorimeter. Set up a ring stand and clamp to suspend a stirring rod and 0.1°C thermometer or temperature probe passing through the lid into the water in the calorimeter. Make sure that the thermometer/probe does not touch the bottom or the sides of the calorimeter.

3. Weigh 4 grams of your assigned hot pack candidate salt to the nearest 0.01 g using an analytical balance, and enter the mass in Table 3.

4. While stirring the water, record the temperature of water for about 4–5 minutes until it has come to thermal equilibrium. Record this initial temperature in Table 3. Add the weighed salt to the water with vigorous mixing. Observe the temperature of the solution every 30 seconds for about 8–10 minutes until the temperature has

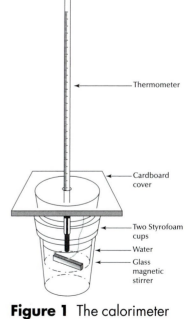

Thermometer

Cardboard cover

Two Styrofoam cups

Water

Glass magnetic stirrer

Figure 1 The calorimeter

stabilized and come to thermal equilibrium. Record this final temperature value in Table 3.

5. Repeat steps 1–3 using 6- and 8-gram samples of the hot pack candidate salt and record your data in Table 3. Calculate ΔT/g for each test result and an average ΔT/g. Does there appear to be a pattern with regard to the amount of salt tested vs. the temperature change?

6. Repeat the entire procedure (steps 1–4) with your assigned cold pack candidate.

TEAM DATA

Table 3 Mass and Temperature Data for 100 mL of Water

Exothermic Salt				Endothermic Salt			
Mass (g)				Mass (g)			
Initial temperature (°C)				Initial temperature (°C)			
Final temperature (°C)				Final temperature (°C)			
ΔT (°C)				ΔT (°C)			
ΔT (°C)/g				ΔT (°C)/g			
Average ΔT *(°C)/g*				*Average* ΔT *(°C)/g*			

7. Do you observe a pattern with regard to the mass of the tested salt and the observed temperature change? Refer to the data to offer an explanation for your conclusion.

8. Collect the different team results for the average ΔT/g for the different salts using a computer or Table 4.

COMPILED TEAM DATA

Table 4 Average $\Delta T/g$ Ratio for Each Salt

Exothermic Salt	Team	Average $\Delta T/g$	Endothermic Salt	Team	Average $\Delta T/g$

Data Analysis and Implications (Part II)

Based on the compiled class data, complete Table 5 indicating the heat absorbed or liberated and the ΔH (heat per mole) for each salt. Calculate the heat absorbed or liberated based on the compiled average $\Delta T/g$ value (Table 4).

Table 5 $\Delta T/g$, Heat Absorbed or Liberated, and ΔH for Each Salt

Exothermic Salt	Average $\Delta T/g$	Heat Released[*]	ΔH (kJ/mol)	Endothermic Salt	Average $\Delta T/g$	Heat Absorbed[*]	ΔH (kJ/mol)

[*]$q = m \times s \times \Delta T$ where $m =$ mass in grams of water
$s = 4.184$ J/g·°C, the same value as that of water.

Note: For all practical purposes, at constant pressure, $q = \Delta H$.

Calculations

4. Put the three endothermic salts in order from maximum to minimum potential as candidates for use in chemical cold packs based on the comparative values for $\Delta T/g$, amount of heat absorbed, and ΔH.

   a. Endothermic salt ranking based on comparative $\Delta T/g$ ratios.

   _____ > _____ > _____

   b. Endothermic salt ranking based on comparative amount of heat absorbed.

   _____ > _____ > _____

Data Analysis and Implications (Part II continued)

    c. Endothermic salt ranking based on comparative ΔH (heat per mole).

         _____ > _____ > _____

    d. Team ranking of candidates for use in chemical cold packs. Justify your final selection and rating of the candidates.

5. Put the three exothermic salts in order from maximum to minimum potential as candidates for use in chemical hot packs based on the comparative values for ΔT/g, amount of heat liberated, and ΔH.

    a. Exothermic salt ranking based on comparative ΔT/g ratios.

         _____ > _____ > _____

    b. Exothermic salt ranking based on comparative amount of heat absorbed.

         _____ > _____ > _____

    c. Exothermic salt ranking based on comparative ΔH (heat per mole).

         _____ > _____ > _____

    d. Team ranking of candidates for use in chemical hot packs. Justify your final selection and rating of the candidates.

Data Analysis and Implications (Part II continued)

6. Based on your data, what observable patterns, if any, are there in the heat behavior of chloride vs. nitrate salts, and +1 cation vs. +2 cation salts?

7. Based on the compiled data, what observable patterns, if any, are there in the heat behavior of salts and salt mass?

Experiment 6–1 Report

Look up the prices of these salts on the Web or in a scientific catalog in order to determine whether the cold pack and hot pack candidates will be economically feasible for the company. Similarly, consider the availability of the salts in the market as another factor for the final selection. Prepare a report of your findings based on the compiled team data and your price research. Make a specific recommendation for a salt to be used in a hot pack and a salt to be used in a cold pack. Indicate the mass of solid salt that should be present in the hot and cold packs.

1. It is a scientific fact that heat energy is required to separate the associated ions in a solid salt, and heat energy is released when water molecules surround (solvate) the ions. $MgSO_4$ is one of the best salts as a possible hot pack. Epsom salt, $MgSO_4 \cdot 7H_2O$, is used for soaking minor sprains (ligaments), strains (tendons), and bruises (as in muscles) and works as a cold pack. How can you explain this phenomenon?

2. Some salts, like silver chloride, $AgCl$, are insoluble in water. What can we conclude, if anything, regarding the energy required for separating and energy released for solvating the ions? Explain.

3. The ΔH per mole of $CaCl_2 = -81.0\,kJ$. Calculate the final temperature of an aqueous solution if 30 g of $CaCl_2$ is added to 100 mL of water at 23°C. Assume that the specific heat of the solution is the same as that of water $= 4.184\,J \cdot g/°C$.

4. Fifty grams of NH_4Cl was dissolved in 100 mL of water. The calculated ΔH for the reaction was $+15.2$ kJ. What is the ΔT obtained in this experiment?

5. A student prepares a cold pack by dissolving 30 grams of ammonium nitrate in 100 g of water at 20°C. If the temperature drops from 20° to 0°C, what is the calculated experimental value for ΔH per mole of ammonium nitrate?

6. Some companies manufacture hot and cold packs that are reusable. How is this accomplished? Check the World Wide Web for information.

Experiment 6-2

How Is Heat Measured Indirectly?

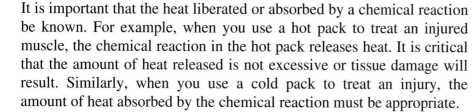

It is important that the heat liberated or absorbed by a chemical reaction be known. For example, when you use a hot pack to treat an injured muscle, the chemical reaction in the hot pack releases heat. It is critical that the amount of heat released is not excessive or tissue damage will result. Similarly, when you use a cold pack to treat an injury, the amount of heat absorbed by the chemical reaction must be appropriate.

Unfortunately, the heat changes associated with many reactions cannot be measured directly. Alternative or indirect methods need to be used. A local manufacturer has asked your chemistry team to present a seminar on indirect methods for measuring heat of chemical reactions. You are to present the heat data results on several reactions as illustrations during your presentation.

- Determine the heat of reactions using a calorimeter.
- Determine the factors that affect the heat of reactions.
- Determine whether the unknown heat of a reaction is linked to the known heat of chemically related reactions.
- Write a report to the manufacturer summarizing the factors that affect reaction heat.

In this investigation, you are to collect heat data and conduct research on factors that affect the heat of a reaction using the following manufacturer-assigned reactions:

(1) $NaOH\ (s) + HCl\ (aq) \rightarrow NaCl\ (aq) + HOH\ (l)$
(2) $NaOH\ (aq) + HCl\ (aq) \rightarrow NaCl\ (aq) + HOH\ (l)$
(3) $NaOH\ (s) \rightarrow NaOH\ (aq)$

You will measure the heat liberated or absorbed by each reaction with a calorimeter.

In this investigation, you will determine the amount of heat exchanged in the dissolution of solid NaOH in hydrochloric acid. For sharing heat data, your instructor will assign teams to different amounts of reactants as indicated in Table 1.

Table 1 Team Assignments, Reaction 1

Team	NaOH (g)	[HCl]
1	2	0.25 M
2	4	0.50 M
3	6	0.75 M

1. The reaction between NaOH (s) and HCl (aq) is exothermic. Will the amount of heat released be different or the same for the different assigned combination amounts of NaOH and HCl? Before starting your heat investigations, record your hypothesis below and then proceed to assemble your calorimeter and make heat measurements.

Hypothesis

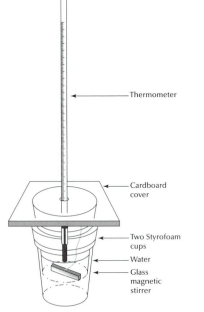

Figure 1 The calorimeter

Thermometer

Cardboard cover

Two Styrofoam cups

Water

Glass magnetic stirrer

Caution: Do not touch NaOH pellets! NaOH can cause serious burns!
Hydrochloric acid can irritate your skin! In case of contact, with NaOH or HCl flush the area with water for several minutes.
Notify your instructor if you spill or contact HCl or NaOH.

---

2. Use two concentric Styrofoam cups with a thermometer/temperature probe and a glass/magnetic stirrer inserted into the cup through a cardboard or plastic coverlid as shown in Figure 1. The thermometer should measure the temperature to $\pm 0.2°C$.

3. Using a graduated cylinder, transfer 200 mL of the assigned concentration of HCl (aq) into the calorimeter. Enter the molarity of HCl into Table 2. Insert a thermometer or probe and a glass or magnetic stirrer to stir the acid. Record the initial HCl temperature in Table 2 once the system comes to thermal equilibrium.

4. Determine and record the mass of the assigned sample of solid sodium hydroxide, NaOH, to the nearest 0.01 g in Table 2. *Note*: It will be difficult to weigh out an exact precalculated mass of NaOH (*s*) since it quickly absorbs moisture from the air and increases in mass. It is therefore essential to weigh the NaOH as quickly as possible to avoid errors in mass measurements. If using standard-size NaOH pellets, 19 pellets weigh about 2 grams.

5. Quickly and carefully, transfer the weighed NaOH into the acid-containing calorimeter. Stir quickly and carefully with the glass or magnetic stirrer and record the highest temperature observed in Table 2. Calculate ΔT, °C (difference between the final and initial temperature), moles of HCl (molarity × volume in liters) and moles of NaOH in the reacting system, and enter the results in Table 2. Dispose of the calorimeter contents by diluting the contents with water unless otherwise specified by your instructor.

Table 2 Team Data: ΔT (°C), NaOH (s) + HCl (aq)

Molarity of HCl	
Moles of HCl	
Mass of NaOH (g)	
Moles of NaOH	
Initial temperature, HCl (°C)	
Final temperature, HCl + NaOH (°C)	
ΔT (°C)	

6. Chemists refer to the amount of heat liberated or absorbed per mole of substance as the molar enthalpy, ΔH, where $\Delta H = q/$(moles of the substance) at constant pressure. Calculate the quantity of heat (q) that is absorbed by the solution and the calorimeter using the information provided below.

INFORMATION

- Use the relationship $q = m \times s \times \Delta T$ for your calculation, where m = mass, s = specific heat, and ΔT (final temperature − initial temperature).
- Assume that the specific heat of the solution is the same as that of water, 4.184 J/g·°C, and that the density of the solution is 1.0 g/mL.
- Assume that any heat exchange between the calorimeter and the surroundings is negligible.

Team Calculations

7. Compare your calculated results with those of any other team using the same amount of reactants. In the event of discrepancies, check for differences in calculations and, if necessary, repeat the heat determination.

8. Share your data with the teams that worked with different amounts of reactants. Compile the data of the different teams using a computer or use Table 3.

Table 3 Heat of Dissolution of NaOH (s) in 200 mL HCl (aq)

Mass NaOH (g)						
ΔT (°C)						
Heat absorbed by solution and the calorimeter (J)						
Moles of HCl						
Heat liberated per mole HCl (kJ/mol)						
Moles NaOH (s)						
Heat liberated per mole NaOH (kJ/mol)						

Data Analysis and Implications (Part I)

1. How does your recorded hypothesis regarding amount of heat released for the different team-assigned amounts of NaOH and HCl compare with your observations and the compiled team data results (Table 3)?

2. Does reagent concentration affect the reaction heat? Support your decision with specific examples from the team-compiled data in Table 3.

3. Does reagent mass affect the reaction heat? For example, if only 2 g of NaOH were used by all teams in combination with the assigned HCl concentration (i.e., 0.25 M, 0.50 M, or 0.75 M HCl), would the heat liberated be the same or different? Why or why not? Support your decision with specific examples from the team-compiled data in Table 3.

4. Does reagent mass or concentration affect the amount of heat liberated per mole? Support your decision with specific examples from the data in Table 3.

5. Calculate and compare the average heat liberated per mole of HCl and of NaOH using the different team results (Table 3).

You will now collect heat data in the same manner as in Part 1 using NaOH(aq) instead of NaOH(s). Different teams will investigate different reactant concentrations (Table 4 to be assigned) and share results.

Table 4 Team Assignments, Reaction 2

Team	[NaOH]	[HCl]
1	0.25 M	0.25 M
2	0.50 M	0.50 M
3	0.75 M	0.75 M

1. Before starting your heat reaction study, record a hypothesis. Do you expect the amount of heat released to be different or the same for the different assigned combination amounts of NaOH(aq) and HCl(aq)? Record your hypothesis below and then proceed to make your heat measurements.

   Hypothesis

2. Transfer 100 mL of the assigned molarity of HCl into the calorimeter. Enter the molarity of HCl in Table 5.

3. Transfer 100 mL of the assigned molarity of NaOH into a separate 250-mL beaker and enter the molarity of NaOH in Table 5.

4. Measure the temperature of each solution once it has come to equilibrium, and record the values as initial temperature values in Table 5. If there is any difference in temperature of the two solutions, record an average value.

5. Pour the solution of NaOH into the solution of HCl. Mix quickly and record the maximum temperature reached as the final temperature in Table 5.

6. Calculate $\Delta T(°C)$, moles of HCl, and moles of NaOH in the reacting system and enter the results in Table 5. Dispose of the calorimeter contents by diluting the contents with water unless otherwise specified by your instructor.

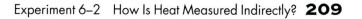

Table 5 ΔT (°C), NaOH (aq) + HCl (aq)

Molarity of HCl	
Moles of HCl	
Molarity of NaOH	
Moles of NaOH	
Initial temperature, HCl (°C)	
Initial temperature, NaOH (°C)	
Final temperature, HCl + NaOH (°C)	
ΔT (°C)	

7. Calculate the heat released per mole of water produced as a result of the reaction between NaOH and HCl and compare your result with any other team using the same quantities of reactants. If necessary, repeat tests to resolve differences.

TEAM CALCULATIONS

8. In Table 6, summarize your team heat results for the reactions from Parts I and II.

Table 6 Molar Enthalpies of Reactions (kJ/mol)

Part I ΔH_1	Part II ΔH_2

9. When you are done, collect the different team data for heat of reaction and ΔH, and ΔH_2 values for the reactions investigated in Part I and Part II in Tables 7 and 8.

COMPILED TEAM DATA

Table 7 Heat of Reactions of 100 mL NaOH (*aq*) + 100 mL HCl (*aq*)—All Teams

Molarity NaOH & HCl	0.25 M	0.25 M	0.50 M	0.50 M	0.75 M	0.75 M
ΔT (°C)						
Heat absorbed by solution and the calorimeter (J)						
Moles of HCl						
Moles NaOH (*s*)						
Heat liberated per mole H_2O (kJ/mol)						

Table 8 Average Molar Enthalpies (kJ/mol)—Parts I and II

Team	1	2	3	4	5	6	Average ΔH
ΔH_1							
ΔH_2							

Data Analysis and Implications (Part II)

6. How does your recorded hypothesis regarding amount of heat released for the different team-assigned concentrations of NaOH and HCl compare with your observations and the compiled team data results in Table 7?

The following chemical reactions represent what you performed in Parts I and II:

(1) $NaOH(s) + HCl(aq) \rightarrow NaCl(aq) + H_2O(l)$ Part I ΔH_1
(2) $NaOH(aq) + HCl(aq) \rightarrow NaCl(aq) + H_2O(l)$ Part II ΔH_2

Based on the ΔH values for reactions 1 and 2, can you predict the unknown heat of the chemically related reaction?

(3) $NaOH(s) \rightarrow NaOH(aq)$ $\Delta H_3 = ?$

Your instructor will assign each team to different amounts (2 g, 4 g, or 6 g) of NaOH for this study, and results will be shared.

1. Record your hypothesis below before starting experimentation. Is the unknown ΔH value for reaction 3 to the ΔH values for the chemically related reactions 1 and 2? When you have finished recording your hypothesis, test your prediction by determining the amount of heat exchanged in the dissolution of solid NaOH.

Hypothesis

2. Pour 200 mL of room-temperature distilled water into the calorimeter. Record the initial temperature in Table 9 once the system comes to thermal equilibrium.

Table 9 ΔT (°C), NaOH (s) → NaOH (aq)

Initial temperature (°C)	
Final temperature (°C)	
Assigned mass of NaOH (g)	
Moles of NaOH	
ΔT (°C)	
Heat per mole NaOH (s)	

Caution: NaOH pellets can cause serious burns! Do not touch the pellets!
In case of NaOH contact, flush the area with water for at least 10 minutes.
Notify your instructor immediately if you contact NaOH.

3. Measure and record the exact mass of the assigned amount of solid NaOH to 0.01 g in Table 9. Remember to be **cautious** with the measurement of the mass of NaOH.

4. Quickly transfer the sample of solid NaOH into the 200 mL of water in the calorimeter. Stir quickly and carefully and record the highest temperature in Table 9 as the final temperature. When you are done, dilute the calorimeter contents by adding the contents to water. Flush the diluted solution down the drain or dispose of as indicated by your instructor.

5. Calculate ΔT (°C), moles of NaOH, and the heat released per mole of NaOH in the reacting system and enter the results in Table 9. Dispose of the calorimeter contents by diluting the contents with water unless otherwise specified by your instructor.

6. Compile the data from teams using different amounts of NaOH (*s*) using a computer or Table 10.

COMPILED TEAM DATA

Table 10 Heat Studies. Reaction 3: NaOH (*s*) → NaOH (*aq*)

Mass NaOH (g)						
ΔT (°C)						
Heat absorbed by solution and the calorimeter (J)						
Moles of NaOH						
Heat liberated per mole NaOH (kJ/mol)						

Data Analysis and Implications (Part III)

7. Based on the compiled team data (Table 10), is the amount of heat liberated related to the mass of solid sodium hydroxide dissolved? If so, how?

8. Does the amount of heat liberated per mole of NaOH depend upon the mass of solid sodium hydroxide dissolved? Refer to the data in Table 10 to support your conclusion.

9. Calculate and record the average team value for heat liberated per mole of NaOH. Compare your predicted value to the actual average value. Explain any differences.

Experiment 6–2 Report

Write a brief report outlining the heat data results to be presented in your seminar on indirect methods for measuring heats of chemical reactions. Be sure to include the relevant heat data results that illustrate factors that affect the heat of a reaction and at least one illustration of measuring an unknown heat of reaction indirectly.

1. This indirect determination of the ΔH value for a reaction has a name. Look for a statement in your text and decide whether the statement provided has any correlation with what you investigated in this experiment.

2. If you had used sulfuric acid, H_2SO_4, instead of HCl, would the experimental ΔH value measured in kJ/mol be different or the same? Explain.

3. Calculate the ΔH for the reaction, $Na^+(g) + Cl^-(g) \rightarrow NaCl(s)$, given the following data:

$$Na(s) + 1/2\,Cl_2(g) \rightarrow Na(g) + Cl(g) \qquad \Delta H° = 230\text{ kJ/mol}$$
$$Na(g) + Cl(g) \rightarrow Na^+(g) + Cl^-(g) \qquad \Delta H° = 147\text{ kJ/mol}$$
$$Na(s) + 1/2\,Cl_2(g) \rightarrow NaCl(s) \qquad \Delta H° = -411\text{ kJ/ml}$$

4. Calculate ΔH in kJ/mol for a neutralization reaction if 50.0 mL of 1.00 M acid, HA, and 50.0 mL of 1.00 M base, BOH, are mixed in a calorimeter. The initial temperature of both solutions was 21.70°C, and the final temperature of the reaction mixture was 27.85°C. Assume that no heat is lost to the surroundings.

5. Assume that the literature value for the heat of neutralization of a strong acid and a strong base is –56 kJ/mol. Use the following hypothetical values of ΔH to predict the value of ΔH for the dissociation of the weak acid HA and the weak base BOH. Write all equations in net ionic form.

$$HA(aq) + NaOH(aq) \rightarrow NaA(aq) + H_2O(l) \qquad \Delta H = -45\text{kJ/mol}$$
$$HCl(aq) + BOH(aq) \rightarrow BCl(aq) + H_2O(l) \qquad \Delta H = -38\text{kJ/mol}$$

Experiment 7–1

Do Like Repel or Attract?

INTRODUCTION

You have accepted a team position with ChemSub in the unit of reagent management. Your first task is to develop tests that will allow you to classify substances present in unlabeled bottles as polar or nonpolar. You then need to refine your tests in order to make distinctions between polar substances relative to degree of polarity.

OBJECTIVES

- Classify liquid samples of known identity based on their interactions with glass, solid surfaces, and other liquids.
- Classify solid samples of known identity based on their interactions (solubility) in liquids.
- Successfully classify unidentified samples as polar or nonpolar based on self-designed tests.

PROCEDURES

In order to develop appropriate tests to successfully classify unidentified substances as polar or nonpolar (Part III), you will qualitatively (Part I) and quantitatively (Part II) study the interactions between some materials and pure chemical substances of known identity. As you research the interactions between these known substances, note any patterns of behavior that might allow you to classify them as polar or nonpolar. For example, some substances are attracted to glass while others are repelled by glass. Do substances with similar behavior (e.g., attraction or repulsion to glass) have like or unlike polarity? Do substances with like properties attract or repel each other?

**PART I—
QUALITATIVE
OBSERVATIONS
OF INTERACTIONS**

Your team is to classify the known materials listed in Table 1 as either polar or nonpolar based on their chemical structures AND their interactions with each other. Before starting your research, observe a demonstration involving two of the materials you will investigate.

Set up two burets. Fill one with water and the other with cyclohexane. Place a beaker under each buret. Rub an inflated balloon or a plastic rod on wool or cloth to produce a charged object. Open the stopcock on the buret containing water. Bring the balloon or other charged object near the stream of water coming from the buret tip and observe the behavior of the liquid. Repeat for cyclohexane. (*Note:* Cyclohexane will dissolve a balloon if allowed to touch it!). Observe and discuss the behavior of the stream of liquid water and cyclohexane in response to the charged object.

1. Record a team hypothesis in Table 1 regarding the classification of each item as either polar or nonpolar *based solely on its chemical structure*.

Table 1 Chemical Structure of Materials

Substance	Structural Formula	Hypothesis: Polar or Nonpolar?
1-Butanol	$CH_3CH_2CH_2CH_2OH$	
Cyclohexane	C_6H_{12}	
Distilled water	H_2O	
Ethanol	CH_3CH_2OH	
Glass	Na^+ and Ca^{2+} salts of the silicate ion, SiO_4^{4-}	
Hexane	C_6H_{14}	
Iodine	I_2	
Methyl orange	$[Na^+(^-O_3S-C_6H_4-N=N-C_6H_4-N(CH_3)_2]$	
Paper towel	$(-CF_2-)n$	
Table salt	$NaCl$	
Potassium nitrate	KNO_3	
Polystyrene	$(C_6H_5-CH=CH_2)_n$	
Sugar	$C_{12}H_{22}O_{11}$	

2. Study the interactions between the materials in Table 1 and chemical substances of known identity using the guidelines provided below. Then develop your own tests and refine your classification of materials to make distinctions between substances that are more or less polar. For example, choose a substance to serve as a screening test (i.e., to search and check out the other materials) to classify the materials as polar or nonpolar based on their solubility interactions. Record all tests and observations and the identity of any screening test material.

GUIDELINES

- Determine whether particular liquids dissolve in one another by adding one drop of liquid at a time to a small amount of the other.
- Determine whether solids dissolve in a liquid by adding several small pieces to ≈ 5 mL of liquid.
- Use a surface as a test. Is the liquid attracted to (does it spread or "wet") or repulsed by (does it bead up) a surface? ⌒ or ⌐⎯ ?

- Use glass as a test. Does a liquid show attraction to, repulsion by, or little interaction with glass? Is the meniscus of the liquid in a glass eyedropper convex, concave, or flat? ⋀ , ⋃ or ⊓ ?

3. When you are done, compare your results with those of another research team and attempt to resolve any differences. If your results do not agree or if your interpretation differs, repeat the tests carefully and see whether you can agree on the observations and their interpretation. What caused the initial difficulty in getting comparable results? What did you do to resolve the differences?

SOLUBILITY TESTS AND OBSERVATIONS

SURFACE TESTS AND OBSERVATIONS

TEAM-DEVELOPED TESTS AND OBSERVATIONS

COMPARISON OF RESULTS

Data Analysis and Implications (Part 1)

1. Did you develop any polar or nonpolar screening test to classify the materials based on their solubility interactions? Indicate your test and the classification of the materials based on their solubility.

2. Did you develop any polar or nonpolar screening test to classify the materials based on their surface interactions? Indicate your test and the classification of the materials based on their surface interactions.

3. Does your classification of materials based on solubility and surface interaction support your hypothesized classification based on chemical structure? Why or why not?

You are required to conduct a minimum of one quantitative investigation (Part IIA or Part IIB) of the interactions of the materials examined in Part I in order to check and refine your classification schemes. You will need a conductivity apparatus for investigation **IIB**. Investigation **IIA** requires a thermometer calibrated in 0.1°C increments.

PART II—QUANTITATIVE OBSERVATIONS OF INTERACTIONS

You and your teammates are to investigate the liquid–liquid interactions of one of three different liquid groups (Table 2).

Part IIA—Temperature

Table 2 Team Assignments: Liquid–Liquid Interactions and Temperature

Group 1	1-Butanol and water	Hexane and pentane	1-Butanol and hexane	Hexane and water
Group 2	Ethanol and water	Cyclohexane and pentane	Cyclohexane and ethanol	Cyclohexane and water
Group 3	Propanol and water	Pentane and hexane	Propanol and pentane	Pentane and I_2 water

Measure out 10 mL of each liquid in your assigned group combination. Pour each liquid into separate clean and dry test tubes and place a thermometer in the first test tube. It is preferable that your thermometer be held in place with a clamp attached to a ring stand since thermometers break easily. Let both tubes stand in a test tube rack for about 5 minutes or until the temperature is constant. Pour the second liquid into the first test tube, and observe and record any temperature change. Repeat the procedure with each liquid pair until consistent (or almost consistent) results are obtained.

When you are done, collect and compare your results with those of other research teams.

Part IIB—Conductivity

Use a conductivity apparatus and water as a test to determine and record the conductivity of water solutions of the materials (water, iodine in water, methyl orange in water, table salt in water, potassium nitrate in water, sugar in water, butanol in water, cyclohexane in water).

When you are done, collect and compare your results with those of other research teams.

TEAM TESTS AND OBSERVATIONS

COMPILED TEAM DATA

Optional Extension   Design experiments to test the interactions of solid and liquid combinations such as iodine and water, iodine and hexane, potassium nitrate and water, potassium nitrate and hexane, methyl orange and water, methyl orange and hexane.

Data Analysis and Implications (Part II)

4. Classify the tested materials and their interactive temperature or conductivity behavior based on the recorded results. Did any samples give results that are difficult to explain?

5. Do your classification results based on temperature or conductivity studies validate prior classifications made in Part I of this experiment? Why or why not?

Conclude your investigation by solving one of the following problems (to be assigned).

A. The labels are missing off reagent bottles containing liquids. Can you identify the most and the least polar liquid?

B. Investigate and classify the following materials:
Moth balls (naphthalene), Parafilm, and tartrazine trisodium salt (FD&C No. 5 for food, drugs, cosmetics, dying wool, and silk)—see *Merck Index* for structure.

C. Investigate a two-phase system of cyclohexane and water:

A 500-mL Erlenmeyer flask contains about 100 mL of cyclohexane and about 100 mL of water. Observe the system. Take a pencil and put black pencil marks (graphite) heavily on one side of a 3 × 5 card. It is preferable to use a soft-lead pencil (#1 or #2). Use an ordinary hole punch to punch out a number of pieces, which will be card stock (paper) on one side and black pencil lead on the other. Add the punched dots to the Erlenmeyer flask and shake. Observe and interpret the result. Identify the liquid in the upper phase based on prior recorded observations. Indicate what you could add to the system (from the tested substances) that would confirm the identity of the liquid on top.

OBSERVATIONS AND ANALYSIS

Add table salt to rubbing alcohol (70% isopropanol, 30% water) until the two phases separate. Suggest a way to determine which phase is which liquid.

Optional Extension

Experiment 7–1 Report

Provide a report summarizing your findings for all investigations. Classify the materials examined (Table 1) as polar or nonpolar, with an indication of substances that may be intermediate in character. If you can, make a scale from nonpolar to highly polar, based on their chemical structures and your studies of interactions between these materials and chemical substances of known identity. For these materials, do materials that are alike (i.e., of the same classification such as polar and polar or nonpolar and nonpolar) repel or attract?

1. Based on your investigation, predict whether Br_2 will be more soluble in tetrachloromethane, CCl_4, or in water. Explain.

2. Predict whether hexane or ethanol will be the best solvent for each of the following and explain why:

   a. KI

   b. CH_3Br

   c. CH_2OH-CH_2OH

3. Carboxylic acids with the general formula $CH_3(CH_2)_nCO_2H$ have a nonpolar $CH_3CH_2 \ldots$ tail and a polar $\ldots CO_2H$ head. Predict the effect of increasing the value of n on the solubility of these acids in polar (such as water) and nonpolar (such as CCl_4) solvents? Explain your reasoning.

4. When an oil spill occurred off the Pacific coastline, special materials had to be used to decontaminate the waters and shoreline. Search the Web for information about oil and oil spills. What must be true of the water solubility, surface interactions, and chemical structure (polar groups present?) of the spilled oils?

5. Methanol, CH_3OH, mixes with water but is insoluble in octane, C_8H_{18}, whereas 1-pentanol, $CH_3CH_2CH_2CH_2CH_2OH$, is soluble in water only to the extent of 2.7 g/100 mL but mixes with octane. Explain this fact.

6. One way of screening potential anesthetics involves testing whether the sample dissolves in olive oil because common anesthetics are soluble in olive oil. Search the Web for information about olive oil and some common anesthetics including cyclopropane (C_3H_6), and halothane ($C_2HBrClF_3$). Why are common anesthetics soluble in olive oil? What other common properties do anesthetics exhibit?

Experiment 8–1

How Can a Chemical Equilibrium Be Shifted?

When a chemical reaction has reached equilibrium, the concentrations of reactants and products remain constant over time. Chemical processing plants often spend much creative energy to determine how to shift equilibria so that desired products are produced in greater quantity and at a reasonable cost. How can the position of an equilibrium be shifted to yield more or less products or reactants? Your team's goal is to identify and explain the conditions that can change the equilibrium position of a reaction.

- Observe the effects of changing conditions on the equilibrium position of a reaction.
- Identify factors that stress an equilibrium system and change the equilibrium position.
- Understand how to alter the equilibrium position of an untested reaction.

Your team will examine the impact of changing concentration and temperature on the position of equilibrium (Part I). Compare your observations with different teams and look for patterns in results to identify factors that shift the position of equilibrium toward reactants or products. In Part II, test your understanding as you make predictions and explain any shift in the equilibrium position of a new equilibrium system.

You and your teammate(s) will observe and interpret the impact of changing conditions (concentration of reactants or products, addition of different reagents, and temperature change) on several chemical equilibrium systems and collect data. When you are done, share your results with the other teams. In case of discrepancies in observations, repeat the procedure.

Part IA—Saturated Sodium Chloride

A saturated solution of sodium chloride, NaCl, is 5.4 M. What conditions will make the precipitation of NaCl (s) as complete as possible? The reaction is:

$$Na^+(aq) + Cl^-(aq) \rightarrow NaCl(s)$$

colorless   white

1. Transfer about 3 mL of saturated NaCl to each of six separate labeled (1–6) test tubes in a test tube rack. Save tube 1 for reference.

2. Take your test tube rack to the fume hood. CAREFULLY add a few drops of concentrated hydrochloric acid, HCl, to the contents of tube 2.

> *Caution:* Keep HCl and HNO$_3$ off your skin and clothing. Wash any contact points with copious amounts of water.

Add concentrated HCl one drop at a time until you have added about 5–6 drops or a change is observed. Compare the contents of this tube with your reference tube. Record the total number of drops added and your observations. Since no additional Na$^+$(aq) ions were added, how do you account for the observations?

3. CAREFULLY add concentrated nitric acid, HNO$_3$, instead of hydrochloric acid, to tube 3. Use the pipet provided in the bottle of acid to add the same number of drops of HCl (step 2). Compare the contents of this tube with the contents of the second tube. Record your observations.

4. Compare the reactants and products from steps 1–3. Which ion is responsible for the observed change in tube 2? How do you know?

5. Use tongs or tweezers to obtain 2 pellets of NaOH. While shaking, CAREFULLY add the 2 pellets of NaOH to tube 4. Mix the contents of the tube with a stirring rod in order to dissolve the pellets.

> *Caution:* Do not touch pellets of NaOH and KOH!
> Keep NaOH and KOH off your skin and clothing!
> Wash any contact points with copious amounts of water.

Record your observations. Since no additional $Cl^-(aq)$ ions were added, how do you account for the results?

6. Use tweezers or tongs to obtain two pellets of potassium hydroxide, KOH. CAREFULLY add the pellets of KOH (instead of NaOH) to tube 5. Mix and compare the contents of this tube with the contents of tube 4. Record your observations.

7. Compare the reactants and products from steps 5–6. Which ion is responsible for the observed changed in tube 5? How do you know?

8. Add acetone one drop at a time using a dropping pipet to tube 6 until about 10 drops have been added or a change is observed. Compare the contents of this tube with the reference tube to which no reagents have been added (tube 1). Record your observations. Since no additional $Na^+(aq)$ ions or $Cl^-(aq)$ ions were added, how do you account for the shift in the equilibrium position of the saturated sodium chloride system?

Part IB—Copper(II) Complex with Water and Hydroxide Ion

Copper(II) ion bonds (complexes) with water and hydroxide ion. What conditions will make the formation of a complex with hydroxide ion as complete as possible? The equilibrium is:

$$[Cu(H_2O)_4]^{2+}(aq) + 2\,OH^-(aq) \rightarrow [Cu(H_2O)_2(OH)_2](s) + 2\,H_2O\,(l)$$

1. Prepare and observe the equilibrium system. Transfer about 0.5 mL (10 drops) of 0.10 M $Cu(NO_3)_2$ to each of three small labeled (1–3) test tubes. Record your visual observations of the properties of the copper(II) complex with water. Add about 0.5 mL (10 drops) of 0.10 M NaOH to each test tube. Record your visual observations of the properties of the resulting copper(II) complex with hydroxide ion. Save tubes 1–3 for reference.

2. Stir, observe, and record your observations while you slowly add 1 mL (20 drops) of 1 M nitric acid, HNO_3, to tube 2. What happened to the copper(II)-hydroxide complex after the addition of acid? Save tube 2 for reference. Record your observations.

3. Stir, observe, and record your observations while you slowly add 1 mL (20 drops) of 1 M sodium nitrate, $NaNO_3$, to tube 3. Record your observations.

4. Compare the reactants used and the products from steps 2 and 3. Which ion is responsible for the observed change in tube 2? How do you know?

5. Now reverse the order of addition of sodium hydroxide and hydrochloric acid to the copper(II) ion previously carried out (tube 2, steps 1 and 2). Transfer 0.5 mL (10 drops) of 0.10 M $Cu(NO_3)_2$ to a clean test tube (tube 4). Add 1 mL (20 drops) of 1 M nitric acid, HNO_3. Stir and add about 0.5 mL (10 drops) of 0.10 M NaOH. Record your observations.

6. Compare the contents of tubes 2 and 4 and your recorded observations. The same reactants (NaOH and HNO_3) were added to the copper(II) ion but in reverse order in tubes 2 and 4. Was the observed change in tube 2 the reverse of that observed in tube 4? Why or why not?

Part IC—Cobalt(II) Complex with Water and Chloride Ion

Cobalt(II) ion bonds (complexes) with water and chloride ion. The conversion of one form of cobalt(II) complex to another involves an energy change that is highly exothermic as the cobalt(II) water complex forms. What conditions will make the formation of a cobalt(II) complex with chloride as complete as possible? The equilibrium is:

$$[CoCl_4]^{2-}(aq) + 6\,H_2O(l) \rightarrow [Co(H_2O)_6]^{2+}(aq) + 4\,Cl^-(aq) + energy$$

blue   pink

1. Add 9–10 crystals of cobalt(II) chloride solid to 20 mL of ethanol in a small beaker until a blue solution results. Add more crystals if needed. Use a pipet or small graduated cylinder to transfer 4 mL of the blue solution to five separate labeled (1–5) test tubes. Save tube 1 for reference.

2. To tube 2 add deionized water. Add one drop at a time until you have added about 10 drops or a change is observed. Record your observations. What happened to the cobalt(II) ions?

3. Add deionized water to tubes 3, 4, and 5 so that tubes 2–5 exhibit the same color. Add about 3–4 small lumps of solid calcium chloride, $CaCl_2$, to tube 3 and 4 and mix to dissolve the solid. Repeat the addition of solid $CaCl_2$ until no more solid dissolves. Record your observations. Since no additional cobalt(II) ions were added, how do you account for the observations?

4. To tube 5, add about 4 small lumps of solid calcium nitrate, $Ca(NO_3)_3$. Record your observations.

5. Compare the reactants and products from steps 3 and 4. Which ion is responsible for the observed change in tubes 3 and 4? How do you know?

6. CAREFULLY add about 1 mL (20 drops) of 0.1 M $AgNO_3$ to tube 4.

> *Caution:* $AgNO_3$ will stain your skin and clothing! Avoid skin contact and wear gloves if available.

   Record your observations. What happened to the cobalt(II) ions?

INFORMATION

- Silver and chloride ions react to form solid silver chloride:
  $$Ag^+(aq) + Cl^-(aq) \rightarrow AgCl(s)$$

7. Place tube 3 in a hot water bath (60°C to 80°C) for 8–10 minutes while shaking once in a while. Record your observations. What happened to the cobalt(II) ions?

8. Chill tube 3 in an ice bath, to see whether the change in step 7 is reversible. Record your observations. Does it appear that temperature is a factor in the equilibrium position of the reaction?

Data Analysis and Implications (Part I)

Part IA—Saturated Sodium Chloride: $Na^+(aq) + Cl^-(aq) \rightarrow NaCl(s)$

1. Did the addition of concentrated HCl to saturated sodium chloride shift the equilibrium toward the reactant (to the left) or product (to the right) side? How do you know?

2. HCl and HNO_3 are both strong acids. There is, however, a difference in the effect of these acids on the saturated sodium chloride equilibrium system (tube 2 vs. tube 3). Explain the observed differences in these two tubes.

3. NaOH and KOH are both strong bases. There is, however, a difference in the effect of these bases on the saturated sodium chloride equilibrium system (tube 4 vs. tube 5). Explain the observed differences in these two tubes.

4. It is a fact that when acetone is added to saturated sodium chloride, the equilibrium shifts toward the right. What observation supports this fact? What is the role of acetone in shifting the equilibrium toward the right side?

Data Analysis and Implications (Part I continued)

5. What reaction conditions appear to shift the position of equilibrium to favor the formation of solid sodium chloride?

Part IB—Copper(II) Complex with Water and Hydroxide Ion:

$$[Cu(H_2O)_4]^{2+}(aq) \rightarrow 2\,OH^-(aq) \rightarrow [Cu(H_2O)_2(OH)_2]\,(s) + 2\,H_2O\,(l)$$

6. How does the addition of 1 M HNO_3 affect the equilibrium? Based on your observations, did the equilibrium shift toward the left or the right side? Explain.

7. When HNO_3 and $NaNO_3$ are independently added to the equilibrium system, there is a difference in observed effect (tube 2 vs. tube 3). Identify the ion causing the equilibrium system to shift toward the left side. What is the role of this ion in shifting the equilibrium toward the left side?

● **Data Analysis and Implications (Part I continued)**

8. Did the order of addition of nitric acid and sodium hydroxide to the equilibrium system make any difference in shifting the equilibrium (tube 2, steps 1 and 2 vs. tube 4 step 5)? Why or why not?

9. What reaction conditions appear to shift the position of equilibrium to make the formation of a complex with hydroxide ion as complete as possible?

● **Part IC—Cobalt(II) Complex with Water and Chloride Ion:**

$$[CoCl_4]^{2-}(aq) + 6\,H_2O\,(l) \rightarrow [Co(H_2O)_6]^{2+}(aq) + 4\,Cl^-(aq) + energy$$

10. Which of the two cobalt(II) complexes was formed after the addition of deionized water (tube 2)? How do you know?

11. When solid calcium chloride was added (tubes 3 and 4), the equilibrium shifted toward the left side. When solid calcium nitrate was added (tube 5) no change in equilibrium position was observed. How do you explain these observations?

12. Based on the information provided, which ion was removed from the equilibrium system upon addition of silver nitrate (tube 4)? Did removal of the ion cause the equilibrium to shift toward the left or right side? How do you know?

13. The equilibrium system of cobalt(II) complex is highly exothermic. Does increasing or decreasing the reaction temperature cause the equilibrium to shift toward the right side (the water complex)? How do you know?

14. Which reaction conditions will make the formation of a cobalt(II) complex with chloride as complete as possible?

Use your acquired knowledge and skills to predict and interpret any change in the position of equilibrium as you alter conditions for a new reaction. Your instructor will assign you to one of the equilibrium systems (Part IIA or Part IIB) below.

The equilibrium system is:

$$2 \, Cu^{2+}(aq) + 4 \, I^-(aq) \rightarrow 2 \, CuI(s) + I_2(s)$$

Part IIA—Copper(II) with Water and Iodide Ion

INFORMATION

- $Cu^{2+}(aq) = [Cu(H_2O)_4]^{2+}$
- Cu^{2+} can react with NH_3 to form the deep blue ammine complex $[Cu(NH_3)_4]^{2+}$.
- The chemical equation given above for the equilibrium system omits spectator species and thus depicts the net reaction.

1. Prepare and observe the equilibrium system: Mix 10 mL of 0.1 M $CuSO_4$ and 10 mL of 0.1 M KI in a small Erlenmeyer flask or beaker.

2. Use your acquired knowledge and skills to explain your observations, including any shift in the equilibrium position and what happens to the copper ion upon addition of the following:

   a. Add 10 mL of hexane to the contents. Mix the contents thoroughly.

   b. Using a dropper, add 5 M NH_3 to the resulting mixture contents of step 2a. Mix the contents thoroughly and add 5 M NH_3 until 3 mL (60 drops) have been added.

   > *Caution:* NH_3 is irritating to the lungs. Perform this test in the hood.

   c. Using a dropper, cautiously add $\sim$5 mL ($\sim$100 drops) of 5 M HCl to the resulting mixture contents of step 2b. Mix the contents thoroughly.

Observations

Part IIB—Copper(II) with Water and Chloride Ion

The equilibrium system is:

$$[Cu(H_2O)_4]^{2+}(aq) + 4\,Cl^-(aq) \rightarrow [CuCl_4]^{2+}(aq) + 4\,H_2O\,(l)$$

1. Observe the equilibrium system: Transfer $\sim$2 mL ($\sim$40 drops) of 1 M copper(II) chloride, $CuCl_2$, to each of two test tubes. Keep one of the two test tubes as a reference as you follow the procedure (step 2) and add the indicated reagents to the second tube. Note that the equilibrium system contains both the blue copper(II) aquo complex, $[Cu(H_2O)_4]^{2+}$, and the yellow-green copper(II) chloro complex, $[CuCl_4]^{2+}$.

2. Use your acquired knowledge and skills to explain your observations including any shift in the equilibrium position and what happens to the copper ion upon addition of the following:

   a. Using a dropper, add $\sim$2 mL ($\sim$40 drops) of ethanol to the contents. Mix the contents thoroughly.

   b. Add 1 mL of deionized water to the resulting mixture contents of step 2a. Mix thoroughly.

   c. Using a dropper, cautiously add $\sim$1 mL ($\sim$20 drops) of 5 M HCl to the resulting mixture contents of step 2b. Mix thoroughly.

   d. Using a dropper, carefully add $\sim$1 mL ($\sim$20 drops) of 0.1 M silver nitrate, $AgNO_3$, to the resulting mixture contents of step 2c. Mix thoroughly.

   > *Caution:* AgNO$_3$ will stain your skin and clothing! Avoid skin contact.

OBSERVATIONS

Data Analysis and Implications (Part II)

Equilibrium System: _____

Explain your observations, including any shift in the equilibrium position. Indicate what happened to the copper(II) ion as you conducted the procedure.

Experiment 8–1 Report

A local chemical processing plant is producing the compounds $NaCl\,(s)$, $[Cu(H_2O)_2(OH)_2]\,(s)$, and $[Cu(H_2O)_4]Cl_2\,(s)$ investigated in this inquiry. What conditions will cause an increase (shift to the right) or decrease (shift to the left) in product yield? Provide examples from your experiment data to support your answer.

1. Search the Web or chemistry reference books for information on Henry Louis LeChatelier and LeChatelier's principle. Which chemical system was studied in his research when he developed LeChatelier's principle? How does LeChatelier's research apply to this experiment?

2. Sodium chloride, NaCl, has a solubility of about 36 g/100mL of water. Thus if 50 g of NaCl is added to 100 mL of water, 36 g will dissolve as Na^+ and Cl^- ions, but 14 g of solid NaCl will remain. Explain how this system appears static but in fact is a dynamic equilibrium, $Na^+(aq) + Cl^-(aq) \rightarrow NaCl(s)$, that is in constant motion though the mass of solid is constant.

3. Pink crystals of the cobalt(II) chloride compound $CoCl_2 \cdot 6H_2O$ actually are $[Co(H_2O)_6]Cl_2$ and are pink because they contain the pink complex $[Co(H_2O)_6]^{2+}$. The $[CoCl_4]^{2-}$ ion is blue. Devices that indicate humidity levels by changing color involve the reaction:

$$2\,[Co(H_2O)_6]Cl_2\,(s) \rightarrow Co[CoCl_4]\,(s) + 12\,H_2O\,(g)$$
$$\text{pink} \qquad\qquad\qquad \text{blue}$$

   a. When the humidity is low, will the equilibrium shift to the right or left? What color will the cobalt(II) crystals be when the humidity is low?

   b. Cobalt(II) crystals are often placed in sealed containers used to store samples to ensure a dry environment. As the crystals remove moisture from the air and samples within the container they change color. Are crystals of the aquo or chloro complex of cobalt(II) used? What color change do you observe as the crystals remove moisture from the air?

4. Ammonia, NH_3, is produced industrially using the equilibrium system below:

$$N_2\,(g) + 3\,H_2\,(g) \rightarrow 2\,NH_3\,(g) + \text{energy}$$

   a. What conditions will optimize the product NH_3?

   b. Check the Web for information on the industrial process (the Haber process) used to produce NH_3. Compare the industrial conditions with your suggestions to optimize NH_3 production.

5. Below is the equilibrium system for an acid-base indicator with a yellow and red form:

$$B\,(aq) + H_2O\,(l) \rightarrow BH^+(aq) + OH^-(aq)$$
$$\text{yellow} \qquad\qquad\quad \text{red}$$

Predict the color of the indicator if you add a few drops of it to:

   a. HCl (aq)

   b. NaOH (aq)

Experiment 8-2

How Do Soaps Compare in Controlling Malodor?

INTRODUCTION

A local company wants to manufacture soap that will control human malodor. One of the characteristics of a good soap that limits malodor is one with minimal quantities of iron(III) metal ions. It has been found that iron(III) metal ions help the underarm bacteria to produce the malodor. You have accepted a research position in this company. Your present task is to determine the amount of iron(III) in several different soaps that are candidates to be marketed as soap to control malodor.

OBJECTIVES

- Investigate and understand the equilibrium reaction system producing a colored iron(III) complex from the reaction of iron(III) nitrate and potassium thiocyanate.
- Construct a calibration curve using known concentrations of iron(III) samples and a spectrophotometer.
- Successfully determine the concentration of iron(III) metal ions in soap samples based on calibration curve results and/or the equilibrium constant of the iron(III) complex.

PROCEDURES

You will determine the concentration of iron(III) ions in samples by spectroscopic analysis. This analysis is possible since iron(III) ions in samples will react to form a colored complex, iron(III) thiocyanate, $[Fe(SCN)]^{2+}$, that absorbs light at 450 nm. A plot of absorbance at 450 nm versus known concentrations of $[Fe(SCN)]^{2+}$ is called a calibration curve. Information from the calibration curve will be used to estimate the iron(III) concentrations in the different soap samples (Parts II and IIIA). A more accurate determination of the iron(III) concentrations can be made using the equilibrium constant for the colored iron(III) thiocyanate complex (Part IIIB).

Your investigation starts with a qualitative study of the reaction system producing the colored iron(III) complex.

1. Add 5 drops of 0.15 M iron(III) nitrate to a small test tube or well plate. Add 5 drops of 0.0015 M potassium thiocyanate, KSCN, and stir, and then record your observations of the resulting mixture.

2. In another tube or well add 5 drops of 0.15 M iron(III) nitrate, $Fe(NO_3)_3$, to 5 drops of 0.0015 M potassium nitrate, KNO_3. Describe and record the appearance of the individual solutions and the resulting mixture.

3. In another tube or well add 5 drops of 0.15 M iron(II) nitrate, $Fe(NO_3)_2$, to 5 drops of 0.0015 M potassium thiocyanate, KSCN. Describe and record the appearance of the individual solutions and the resulting mixture.

4. Compare the reactants and results from steps 1–3. Which ion is needed for the color change occurring in the reaction under investigation (step 1)? How do you know?

5. Assuming the reaction involves the coming together of ions in solution, which ions reacted to cause the observed color change (step 1)? Consult the *CRC Handbook of Chemistry and Physics* (Inorganic Compounds section) and record the published characteristics of possible products. Do the published properties verify your choice of reacting ion combinations?

6. Place 25 mL of 0.0015 M potassium thiocyanate, KSCN, and 25 mL of water into a small beaker. Put 3 or 4 drops of 0.15 M iron(III) nitrate, $Fe(NO_3)_3$, into the beaker and stir the mixture. Divide the solution mixture into four labeled test tubes.

7. Save the first tube as a reference. To the second tube add some crystals of KSCN. Compare and record the color of the second and first tube. Does the mixture (tube 2) now contain more or less of the colored iron(III) complex? Since no more iron(III) ions were added, how do you account for the color change?

8. To the third tube add 5 drops of 0.15 M iron(III) nitrate, $Fe(NO_3)_3$. Compare and record the color of the third and first tube. Does the mixture (tube 3) now contain more or less of the iron(III) colored complex? Since no more SCN^- ions were added, how do you account for the color change?

9. To the fourth tube add 5 drops of 0.10 M NaOH to form $Fe(OH)_3 (s)$. Compare and record the color of the fourth and first tube. Does the mixture now contain more or less of the iron(III) colored complex? Since SCN^- ions were not removed, how do you account for the color change?

Date: _____ Name: _____ Class: _____

Data Analysis and Implications (Part I)

1. What do you observe when you mix KSCN and $Fe(NO_3)_3$? What is the evidence that a chemical reaction occurred?

2. Thiocyanate (SCN^-) ion is used as an indicator for the iron(III) ion. How? Refer to your observations.

3. What are the spectator ions in the reaction between KSCN and $Fe(NO_3)_3$? How do you know?

4. Write a balanced net equation that represents the reaction between iron(III) nitrate and potassium thiocyanate and is consistent with the data.

Data Analysis and Implications (Part I continued)

5. When additional KSCN (*s*) was added to the reaction mixture, what ion had to be present in the reaction mixture to account for your observations?

6. When additional iron(III) nitrate is added to the reaction mixture, what ion had to be present in the reaction mixture to account for your observed results? Note that no additional SCN^- ions were added. Explain this observation.

7. Before adding additional reactants (potassium thiocyanate or iron(III) nitrate), were SCN^- and/or iron(III) ions present in the solution? Explain why or why not. Refer to your observations to support your conclusion.

The research teams will prepare calibration curves (a plot of absorbance vs. concentration) at 450 nm from absorbance readings of solutions of different concentrations of $FeSCN^{2+}$ (*aq*). The calibration curve information will be used to estimate the iron(III) concentrations in different soaps (Part III). Teams will use assigned concentrations of the reactants KSCN and $Fe(NO_3)_3$ (Table 1) to prepare the different concentrations of $FeSCN^{2+}$ (*aq*).

INFORMATION

- For solution preparation you will need the following: two 50-ml burets, a ring stand and clamps, five clean and dry 50-mL beakers, three clean and dry 100-mL beakers, and a 5-mL pipet.

Table 1 Team Reactant Concentrations for $FeSCN^{2+}$ (*aq*) Preparation

Team	KSCN (M)	$Fe(NO_3)_3$ (M)
1	0.0025	0.25
2	0.0020	0.20
3	0.0015	0.15

> *Caution:* Potassium thiocyanate (KSCN) is highly toxic. Do not touch!
> The metal ion solutions are toxic. Do not dump the metal ion solutions down the sink.
> Discard all waste in the appropriate waste container.

1. Obtain about 50 mL of $Fe(NO_3)_3$ at the assigned team concentration in a clean and labeled 100-mL beaker. Label five clean and dry 50-mL beakers from 1 to 5. Transfer 5 mL of the $Fe(NO_3)_3$ solution with a pipet to each beaker.

2. Transfer the assigned team concentration of KSCN to one buret and transfer 0.10 M HNO_3 to the other buret. Prepare the five standard solutions by delivering the quantities of KSCN and HNO_3 indicated in Table 2 into the 5-ml samples of iron(III) nitrate (beakers 1–5).

3. Follow the instructions to calibrate and use the spectrophotometer to determine the absorbance of each sample at 450 nm. Use the 0.10 M HNO_3 as the reference sample to calibrate (set absorbance equal to 0 or transmittance equal to 100%) the spectrophotometer at 450 nm. Measure and record the absorbance of each standard solution at 450 nm in Table 2. Compare your results with those of any team

using the same solution concentrations. In the event of differences, repeat your measurements.

TEAM DATA

Table 2 Molarity and Volumes of KSCN and $Fe(NO_3)_3$ versus Absorbance at 450 nm

Solution	____M KSCN (mL)	____M $Fe(NO_3)_3$ (mL)	0.1 M HNO_3 (mL)	Absorbance
1	5.00	5.00	15.00	
2	4.00	5.00	16.00	
3	3.00	5.00	17.00	
4	2.00	5.00	18.00	
5	1.00	5.00	19.00	

4. To plot the team calibration curve, calculate the concentration of $[FeSCN]^{2+}$ in each sample and enter your results in Table 3. Note that you will need to do the following:

   a. Determine the number of moles of KSCN and $Fe(NO_3)_3$ in each sample.

   b. Determine the concentration, molarity, of the reactants SCN^- and Fe^{3+}, in each sample.

   c. Determine the limiting reagent and calculate the concentration of $[FeSCN]^{2+}$ for each sample.

CALCULATIONS

Table 3 Calibration Curve Data

Solution #	Moles KSCN	Moles $Fe(NO_3)_3$	$[SCN^-]$ (M)	$[Fe^{3+}]$ (M)	$[Fe(SCN)]^{2+}$ (M)	Absorbance
1						
2						
3						
4						
5						

5. Collect and compile the team data from groups using the same reactant concentrations using a computer or Table 4. Calculate the average values.

CALCULATIONS

Table 4 Compiled Team Data for Calibration Curve

Solution #	Moles KSCN	Moles Fe(NO$_3$)$_3$	[SCN$^-$] (M)	[Fe^{3+}] (M)	[FeSCN]$^{2+}$ (M)	Absorbance
1						
Average						
2						
Average						
3						
Average						
4						
Average						
5						
Average						

- ## Data Analysis and Implications (Part II)

8. Prepare and attach a calibration curve of absorbance vs. concentration of $[FeSCN]^{2+}$ using the compiled team data (Table 4). Determine and record the slope of the best straight line through the points.

9. Determine k, the proportionality factor between the absorbance and the concentration. Beer's law indicates that $A = \varepsilon l c$, where A = absorbance, ε = molar absorptivity, l = path length, and c = concentration in molarity. The proportionality constant, k, is typically equal to εl.

   Note: The proportionality factor k is the slope of the best straight line through the points of your calibration curve. You will use this information to estimate the unknown iron(III) ion concentrations in soap samples (Part III).

10. How do your k values compare with those of other teams using the same or different reactant concentrations? Note any differences or similarities in the results.

11. Calculate the average k values based on the compiled team data.

You will be provided with samples of different soap solutions. Your task is to determine and compare the amount of iron(III) in the different soaps. You can get a good estimate of the concentration of iron(III) by spectroscopic analysis (Part IIIA) using the calibration curve prepared in Part II. A more accurate determination of the iron(III) concentrations can be made using the equilibrium constant value (Part IIIB) for the reaction producing the iron(III) thiocyanate complex.

Estimate and compare the concentration of iron(III) in several commercially available soap extract samples. After reading the guidelines provided below, record your procedure, observations, and results.

Part IIIA—Calibration Curve Estimation

INFORMATION

- Preliminary analysis of soap samples indicates that the concentrations of iron(III) ions are between 0.0010 M and 0.0030 M.

GUIDELINES

- Convert iron(III) ions in the soap sample solutions to the iron(III) thiocyanate complex ion by adding KSCN to the soap samples.
- KSCN (*aq*) of different concentrations (0.0015 M, 0.0020 M, 0.0025 M) will be available on the reagent shelf.
- Using your calibration curve (Part II) and the average value of *k* based on teams using the same reactant concentrations, determine the concentration of the iron(III) thiocyanate complex in each sample.
- Estimate the concentration of iron(III) ions in each soap sample solution based on the concentration of iron(III) complex in each soap sample.

PROCEDURE AND OBSERVATIONS

Soap samples _____, _____, _____

Estimated iron(III) ion concentrations:

Part IIIB—Equilibrium Constant Determination

Determine the concentration of iron(III) ions in the different soap sample solutions based on the equilibrium constant, K_c, for the reaction being studied:

$$\text{Net reaction} : Fe^{3+}(aq) + SCN^-(aq) \rightleftarrows [Fe(SCN)]^{2+}(aq)$$

$$K_c = \frac{[Fe(SCN)]^{2+}}{[Fe]^{3+}[SCN^-]}$$

Search the Web or use the *CRC Handbook* or another reference text to obtain the K_c value at room temperature. Calculate and compare the concentration (M) of iron(III) ion in the different soap sample solutions using K_c. Record your calculations and results below.

CALCULATIONS AND RESULTS

$K_c =$ _____ Reference source: _____

● **Data Analysis and Implications (Part III)**

12. What is the concentration of iron(III) ion in the soap sample solutions based on:

    a. Calibration curve estimation results (Part IIIA)

    Use the average value of k based on teams using the same reactant concentrations, for determinations of your concentrations. Record all calculations.

●

    b. Equilibrium constant determination results (Part IIIB)

    Record the value of K_c and all calculations.

●

Experiment 8–2 Report

Based on your investigation, which soap sample should the company manufacture to control malodor? Why? Refer to your data, and give specific examples of results to support your recommendation.

1. Search the Web or other text resources for information on causes of malodor and the presence of iron(III) ions in soap.

Questions 2–5 refer to the data below obtained when a student reacted KSCN (aq) and $Fe(NO_3)_3$ (aq) at the indicated concentrations and measured the absorbance of the product mixture at 450 nm.

#	$[SCN^-]_{initial}$	$[Fe^{3+}]_{initial}$	Absorbance	$[Fe(SCN)]^{2+}_{eq}$	$[Fe^{3+}]_{eq}$	$[SCN^-]_{eq}$	K_c
1	0.000500	0.0500	0.770	0.000479	0.0495	0.0000211	458
2	0.000500	0.0250	0.683	0.000425	0.0246	0.0000752	230
3	0.000500	0.0125	0.623	0.000387	0.0121	0.000113	283
4	0.000500	0.00625	0.511	0.000318	0.00593	0.000182	295
5	0.000500	0.00313	0.386	0.000240	0.00289	0.000260	319
6	0.000500	0.00156	0.258	0.000160	0.00140	0.000340	336

2. In which sample is the formation of the iron(III) complex ion the most complete? How do you know?

3. Which sample is likely to have the most accurate equilibrium constant value? Why?

4. Based on the absorbance value, describe what happens on a molecular level as $[Fe^{3+}]$ is increased.

5. If 5 mL of each of the reactants of sample 4 is mixed and diluted by a factor of two by adding 10 mL of dilute nitric acid, the sample absorbance decreases by a factor of four rather than a factor of two. The dilutions and absorbance readings are not in error. Offer an explanation for this observation.

Experiment 9-1

Is It Acidic, Basic, or Neutral?

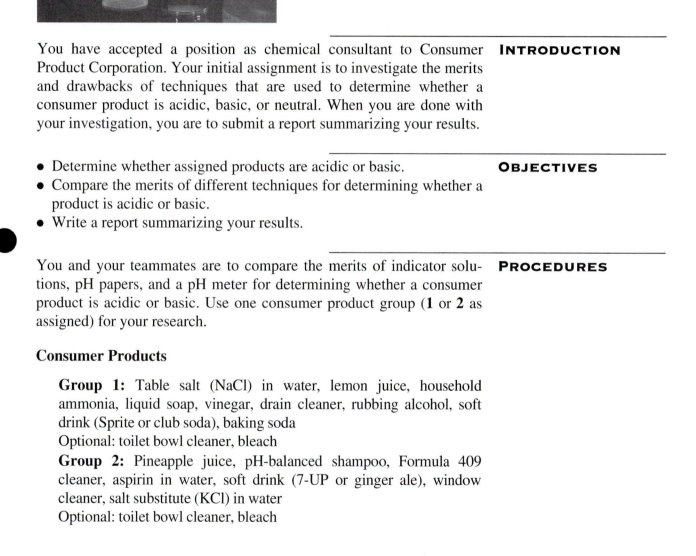

You have accepted a position as chemical consultant to Consumer Product Corporation. Your initial assignment is to investigate the merits and drawbacks of techniques that are used to determine whether a consumer product is acidic, basic, or neutral. When you are done with your investigation, you are to submit a report summarizing your results.

INTRODUCTION

- Determine whether assigned products are acidic or basic.
- Compare the merits of different techniques for determining whether a product is acidic or basic.
- Write a report summarizing your results.

OBJECTIVES

You and your teammates are to compare the merits of indicator solutions, pH papers, and a pH meter for determining whether a consumer product is acidic or basic. Use one consumer product group (**1** or **2** as assigned) for your research.

PROCEDURES

Consumer Products

Group 1: Table salt (NaCl) in water, lemon juice, household ammonia, liquid soap, vinegar, drain cleaner, rubbing alcohol, soft drink (Sprite or club soda), baking soda
Optional: toilet bowl cleaner, bleach

Group 2: Pineapple juice, pH-balanced shampoo, Formula 409 cleaner, aspirin in water, soft drink (7-UP or ginger ale), window cleaner, salt substitute (KCl) in water
Optional: toilet bowl cleaner, bleach

Consumer Product Corporation asks that you measure the pH of each consumer product with both pH paper and a pH meter or, if available, a pH probe and a CBL.

1. Before taking any pH measurements, prepare a table (Table 1) in the space provided for recording your pH readings. Include separate columns for pH measurements of each consumer product with pH paper and the pH meter.

2. Follow the guidelines below for conducting your pH studies. When you are done, share and compare your pH results with a different team investigating the same group of consumer products. In the event of discrepancies, repeat your pH measurements.

INFORMATION

- $pH = -\log [H^+]$; pH 7 is neutral, pH <7 is acidic, pH >7 is basic.
- pH paper comes in broad-range (such as pH 1–11) or narrow- (such as pH 1–3) range forms.
- Product samples may be placed in a small beaker for the purpose of taking pH readings.
- You will need approximately 30 mL of each sample for taking pH measurements.

GUIDELINES

- Do *not* immerse pH paper in the product sample! Dip a *clean* stirring rod into the sample and touch the wet rod to the paper. To determine the pH of each sample, match the color of the wet pH paper to the provided pH paper color code.
- Take an initial pH paper reading with broad-range acid (pH 1–7) or base (pH 7–14) paper as appropriate. Based on the result (e.g., acid product reads $\cong$ pH 2) choose a narrow-range pH paper (e.g., pH 1–3) to determine the product sample's actual pH.
- Avoid errors in pH meter readings! Calibrate the pH meter with an acid (base) buffer when taking the pH of acid (base) samples! Use a buffer with a pH that is close to the pH of your consumer product sample ($\pm$1 pH unit).

Procedure and Observations

Table 1 Team Data: Consumer Product pH   Group _____

COMPARISON OF RESULTS

Data Analysis and Implications (Part I)

1. Based on pH measurements, which consumer products are acidic? List the products and compare their degrees of acidity.

2. Based on pH measurements, which consumer products are basic? List the products and compare their degrees of basic character.

3. Are the pH paper and pH meter measurements in agreement for the different consumer products as acids or bases? Indicate any difference in results.

Consumer Product Corporation asks that you now test the pH of your assigned consumer products using indicators. The indicators you will use are phenolphthalein, bromthymol blue, and bromphenol blue (or as assigned).

1. Create a table (Table 2) for recording your data in the space provided under Team Data. Include a column to record observations of the product color itself and separate columns for color change results with each indicator, as well as a column for observations of the reference blank (see Guidelines below).

2. Follow the guidelines below to conduct your product tests. Make sure to look for any color changes in the product solution upon addition of the indicator. To determine whether there is a color change, it is best to keep a sample of the product solution sample alone for comparison purposes. If there is no color change—i.e., the color upon addition of the indicator is the same as the original product color—record "no change." If the product color (e.g., "blue") is different upon addition of the indicator, record the color change (e.g., "darker blue").

GUIDELINES

- To minimize waste, test a few drops of the consumer product with a few drops of the indicator.
- Use a spot plate or well plate (preferably white) or small test tubes for testing products.
- The reference blank can be a duplicate sample of the product to which you add no indicator or a water sample to which you add indicator.
- All materials used in small quantities can be flushed down the drain with water.
- Compare your team results with those of other teams testing the same products in order to detect any errors. If necessary, repeat your tests.

> *Caution:* Some products are potential eye irritants.
> Read the labels on all products. Adhere to any safety guidelines.
> If a product contacts your skin or clothes, notify your supervisor.

TEAM DATA

Table 2 Indicator Studies: Consumer Products   Group _____

COMPARISON OF RESULTS

Date: _____ Name: _____ Class: _____

Data Analysis and Implications (Part II)

4. Is there consistency and relationship in the color changes of the different indicators? For example, does the appearance of a particular color of bromthymol blue indicator appear to correspond to the appearance of some particular color of other tested indicators? List the indicators and colors that correlate.

5. List the tested indicators. What color is each indicator in an acidic solution? What color is each indicator in a basic solution?

6. Do any indicators give conflicting results for a particular consumer product? If so, list the indicators and the results.

7. List any consumer product and indicator test details where the pH measurement (Part I) conflicts with indicator results.

Consumer Product Corporation asks, does product concentration affect determination of acid–base characteristics? Several of the tested consumer products contain hydrochloric acid, HCl, or sodium hydroxide, NaOH. You are to investigate the acid–base characteristics of a range of concentrations of HCl and NaOH using both pH paper and the indicator solutions previously used in Part II.

1. Prepare a table (Table 3) in the space provided under Team Data for recording your data. Include separate rows for six different concentrations of HCl and of NaOH and separate columns for the pH data from the pH paper and the different indicator solutions.

2. Follow the procedure below for preparing six different concentrations of HCl and NaOH and obtaining pH data.

   > *Caution:* Do not allow acids or bases to contact your body. If contact occurs, wash the area profusely with cold water. Notify your instructor of any spills or body contact with acids or bases.

   a. Obtain 10 mL of 0.1 M HCl in a *clean* and *dry* test tube labeled "10^{-1} M HCl."

   b. Prepare 0.1 M to 0.000001 M (10^{-1} M to 10^{-6} M) concentrations of HCl via serial dilution.
   Use a clean disposable pipet to transfer and measure acid for dilutions. Use a *clean graduated* cylinder or buret to transfer and measure water.

   Prepare 10^{-2} M HCl by transferring 1 mL (20 drops) of the 10^{-1} M HCl and 9 mL of distilled water to a second *clean* and *dry* test tube. Mix, and label the second test tube "10^{-2} M HCl." Transfer 1 mL (20 drops) of the 10^{-2} M HCl sample to a third test tube, add 9 mL of distilled water, and mix. Label the third tube "10^{-3} M HCl." Continue this serial dilution until you have 6 samples of HCl from 10^{-1} M to 10^{-6} M.

   c. Prepare 10^{-1} M to 10^{-6} M concentrations of NaOH via serial dilution. Obtain 10 mL of 0.1 M NaOH and carry out a serial dilution similar to step 1. When you are done you should have 6 solutions labeled from "10^{-1} M NaOH" to "10^{-6} M NaOH."

3. Determine the pH of each sample using both pH paper and indicator solutions. Test a few drops of each sample concentration with a few drops of each indicator previously tested in Part II. Record your pH data in Table 3.

Table 3 Acid/Base Concentration and pH Paper vs. pH Indicators

Data Analysis and Implications (Part III)

8. Do any indicators give conflicting results as the concentration of the sample is altered? List any indicator that changes color as the concentration of the hydrochloric acid or sodium hydroxide sample is altered. Indicate the concentration of acid or base where the color change occurs.

9. Do the pH measurements relate to one another as the concentration of HCl is altered? How?

10. Do the pH measurements relate to one another as the concentration of NaOH is altered? How?

11. Are the different indicator and pH results in agreement as HCl and NaOH concentration alters? List any indicator and pH range conflicts.

Date: _____ Name: _____ Class: _____

Experiment 9–1 Report

Summarize your findings regarding the merits and drawbacks of indicator solutions, pH papers, and a pH meter for determining whether a compound is acidic or basic. Refer to your observations and analysis of results to support your conclusions.

1. A bottle of inexpensive white wine has a pH of 2.85.

   a. What color will the indicator phenolphthalein display when added to the white wine?
   b. What is the hydrogen ion concentration, $[H^+]$, of the white wine?
   c. Is the white wine acidic or basic?

2. The label on a shampoo bottle reads: "For Professional Use Only. Acid-balanced conditioning shampoo, which enhances hair structure—pH 5."

   a. What is the hydrogen ion concentration, $[H^+]$, of the shampoo?
   b. Is the shampoo acidic or basic?
   c. Why do you think the company has put "pH" and "acid-balanced" on the label?

3. The pH of a cup of coffee at $25°C$ is 4.2. What is the $[H^+]$ in the coffee?

4. A solution of 0.100 M HCl contains phenolphthalein indicator and is colorless. A solution of 0.100 M NaOH contains phenolphthalein indicator and is pink in color. What will the color be if equal volumes of the two solutions containing the indicators are mixed?

5. You are given three burets that contain either 0.10 M NaOH, 0.20 M NaOH, or 0.10 M HCl. How can you identify the contents of each buret if all you have available are flasks and a bottle of bromo-thymol blue liquid indicator?

Experiment 9–2

Are Acid–Base Properties in Consumer Products Predictable?

INTRODUCTION

You continue to enjoy your position as one of a team of chemical consultants to Consumer Product Corporation. Your current assignment is to collect data regarding acid–base properties and the chemical composition of consumer products. Consumer Product Corporation wishes to know whether the acid–base properties of a consumer product are predictable from its chemical composition. When you are done with your research, you are to provide a report summarizing your findings.

OBJECTIVES

- Determine whether acid–base classification of substances is predictable from chemical composition.
- Determine whether acid–base properties, including reactivity, are linked to chemical composition.
- Prepare a scientific report summarizing your results.

PROCEDURES

Acids and bases often contain the elements oxygen and hydrogen in combination with another element. You are to study compounds of the general formula $X_aO_bH_c$, where a, b, and c are integers and X represents any element. In order to collect and validate data quickly, different teams will study acid-base properties of different groups of compounds (to be assigned from Table 1 below) and share results (Parts I and II). Your study concludes (Part III) by asking you to design experiments to identify the acid and/or base content of an unlabeled bottle from a list of possible consumer products.

Table 1 Team Assignments: 0.01 M $X_aO_bH_c$ Compounds*

Group 1	LiOH	BaO_2H_2	CO_2H_2	$C_2O_2H_4$	NO_3H
Group 2	KOH	CaO_2H_2	SeO_4H_2	SO_3H_2	SO_4H_2
Group 3	NaOH	MgO_2H_2	BO_3H_3	PO_4H_3	ClO_4H

*Samples are 0.01 M except for MgO_2H_2 (saturated solution @ 1.6×10^{-4} M).

Consumer Product Corporation wishes to know whether the acid–base character of a substance is predictable from its chemical composition. In Part I you will determine if acid–base character can be predicted from the number of oxygen or hydrogen atoms or from the identity of the element in such compounds.

INFORMATION

- You will need approximately 50 mL of each solution.
- Bromthymol blue indicator is yellow in acidic solution and blue in basic solution.
- pH paper comes in broad-range (such as pH 1–11) or narrow- (such as pH 1–3) range forms.
- To determine the pH of a sample using pH paper, dip a *clean* stirring rod into the sample and touch the wet rod to the paper. Match the color of the wet pH paper to the provided pH paper color code.

1. Does the acid–base character of $X_aO_bH_c$ compounds relate to their chemical composition? Record a hypothesis. For example, do you expect the acid or base properties of a sample to be predictable from its chemical composition? If so, will its acid–base properties depend on the number of H or O atoms, or on the identity of the element X or some other factor?

Hypothesis

2. Determine whether each assigned group sample is an acid or a base. Transfer a small amount of sample (e.g., 1/2 mL or less) to a small test tube or well of a well plate and add one drop of bromthymol blue indicator solution. Record your observations and conclusions.

3. Rewrite the formulas to reflect the acid or base character of each assigned sample.
   Note: Formulas for acids are *usually* written H_nXO_n, and formulas for bases are usually written $X(OH)_n$, where X is an element and n is some integer.

4. Determine and record the pH of your identified acid and base compounds qualitatively using pH paper. Take an initial pH paper reading with broad-range acid (pH 1–7) or base (pH 7–14) paper as appropriate. Based on the result (e.g., acid product reads $\cong$ pH 2) choose a narrow-range pH paper (e.g., pH 1–3) to determine the sample's pH.

5. Determine and record below the pH data for your assigned compounds quantitatively using a pH meter or, if available, a pH probe and CBL. Make sure to calibrate the pH meter with the provided acid (base) buffers when taking the pH of acid (base) samples! Use a buffer with a pH that is close to the pH of your sample (± 1 pH unit).

6. Compare your pH results with those of any other team testing the same group of compounds. If the pH data are comparable, there is no need to take additional pH readings. If the pH data are not comparable, repeat your pH test procedure, until readings are similar. When you are done, collect the different team results using a computer or Table 2.

Table 2 pH of 0.01M $X_aO_bH_c$ Compounds

Compound	pH	Average pH
LiOH		
BaO_2H_2		
CO_2H_2		
$C_2O_2H_4$		
NO_3H		
KOH		
CaO_2H_2		
SeO_4H_2		
SO_3H_2		
SO_4H_2		
NaOH		
MgO_2H_2		
BO_3H_3		
PO_4H_3		
ClO_4H		

Data Analysis and Implications (Part I)

1. Enter the average pH from Table 2 of tested $X_aO_bH_c$ compounds in the periodic table in the position of element X.

1																	18
1 H	2											13					2 He
3 Li	4 Be											5 B	6 C	7 N	8 O	9 F	10 Ne
11 Na	12 Mg	3	4	5	6	7	8	9	10	11	12	13 Al	14 Si	15 P	16 S	17 Cl	18 Ar
19 K	20 Ca	21 Sc	22 Ti	23 V	24 Cr	25 Mn	26 Fe	27 Co	28 Ni	29 Cu	30 Zn	31 Ga	32 Ge	33 As	34 Se	35 Br	36 Kr
37 Rb	38 Sr	39 Y	40 Zr	41 Nb	42 Mo	43 Tc	44 Ru	45 Rh	46 Pd	47 Ag	48 Cd	49 In	50 Sn	51 Sb	52 Te	53 I	54 Xe
55 Cs	56 Ba	57 La	72 Hf	73 Ta	74 W	75 Re	76 Os	77 Ir	78 Pt	79 Au	80 Hg	81 Tl	82 Pb	83 Bi	84 Po	85 At	86 Rn

Are acid–base properties of $X_aO_bH_c$ compounds related to the position of X in the periodic table? If so, how? For example, can you reliably predict the acid–base classification of an untested compound such as AlO_3H_3 or PO_3H_3?

2. Can acid–base properties of $X_aO_bH_c$ compounds be predicted from the electronegativity (attraction for electrons in a bond) value of element X?

    a. Look up and record the electronegativity values of element X in the tested compounds (Table 2) in the *CRC Handbook* or another text.

Data Analysis and Implications (Part I continued)

b. List below the $X_aO_bH_c$ compounds in order of *decreasing* pH vs. the electronegativity of element X. Is there a pattern? For example, can you reliably predict the acid–base classification of an untested compound with the general formula $X_aO_bH_c$ given only the electronegativity of X?

3. Can acid–base properties of $X_aO_bH_c$ compounds be predicted based on the integer c? List all tested compounds in order of increasing number of H atoms per tested compound formula vs. pH. What do the results indicate?

● **Data Analysis and Implications (Part I continued)**

4. Can acid–base properties of $X_aO_bH_c$ compounds be predicted based on the integer b? Organize the compounds based on number of O atoms per tested compound vs. pH. What do the results indicate? For example, compare the pH data for SO_3H_2 and SO_4H_2.

Consumer Product Corporation asks that you determine if acid and base reactivity is linked to the chemical composition of acids and bases. Specifically, you are to study the reactivity of the acids and bases and look for similarities or differences in reactivity behavior. In order to collect extensive data quickly, the different teams are to study the reactivity of solutions of the $X_aO_bH_c$ compounds previously explored in Part I at concentrations of 1.0 M and share results.

1. Obtain a small volume ($\cong 5$ mL) of each assigned compound used in Part I at concentrations of 1.0 M and store them in small separate labeled test tubes or vials or beakers. Record the identity of assigned compounds in column one of Table 3.

2. Based on the average pH of $X_aO_bH_c$ compounds (Table 1) determine and record in Table 3 if each assigned compound is an acid or base.

3. Divide an original 1.0-M solution into three individual wells of a well plate or three individual small tubes for testing reactivity. When you have completed the reactivity tests below, proceed to test all remaining assigned samples in a similar manner.

Reactivity Tests

Place a small piece of magnesium metal (Mg) into the first sample, 3–5 drops of 1 M magnesium nitrate ($Mg(NO_3)_2$) into the second sample, and a single chip of marble ($CaCO_3$) into the third sample. Record all observations in Table 3. Share and compare team results with any team testing identical samples. In the event of any discrepant results, repeat the tests.

TEAM DATA

Table 3 Acid–Base Reactivity of $X_aO_bH_c$ Compounds

Compound	Acid or Base?	Mg Metal	Mg^{2+} Ion	$CaCO_3$

4. Collect the results of the different teams using a computer or Table 4.

COMPILED TEAM DATA

Table 4 Acid–Base Reactivity of $X_aO_bH_c$ Compounds

Compound	Mg Metal	Mg^{2+} Ion	$CaCO_3$
LiOH			
BaO_2H_2			
CO_2H_2			
$C_2O_2H_4$			
NO_3H			
KOH			
CaO_2H_2			
SeO_4H_2			
SO_3H_2			
SO_4H_2			
NaOH			
MgO_2H_2			
BO_3H_3			
PO_4H_3			
ClO_4H			

Data Analysis and Implications (Part II)

5. Based on the average pH of $X_aO_bH_c$ compounds (Table 1) identify the compounds that are acids. Do the acid compounds show consistent reactivity behavior with Mg metal, Mg^{2+} ion, or $CaCO_3$ (Table 4)?

Write a chemical equation for any one of the reactions you observed with Mg metal, Mg^{2+} ion, or $CaCO_3$ to represent the observed reaction or write "no reaction."

Mg metal:

Mg^{2+} ion:

$CaCO_3$:

6. All the $X_aO_bH_c$ compounds that are acids were tested at the same concentration (1.0 M). Since they are at the same concentration we might expect all to have the same amount of whatever it is that makes them acids. Do your observations indicate that equal concentrations of the different acids have equal or different "strengths" (i.e., equal or different reactivity)? Refer to specific observations to support your conclusion.

7. Based on the average pH of $X_aO_bH_c$ compounds (Table 1) identify the compounds that are bases. Do the base compounds show consistent reactivity behavior with Mg metal, Mg^{2+} ion, or $CaCO_3$ (Table 4)?

Write a chemical equation for any one of the reactions you observed with Mg metal, Mg^{2+} ion, or $CaCO_3$ to represent the observed reaction or write "no reaction."

Mg metal:

Mg^{2+} ion:

$CaCO_3$:

Consumer Product Corporation asks that you test what you have learned from your study. Your supervisor will assign you to *one* of the following problems.

1. Determine the contents of bottles containing solid samples of consumer products. The consumer products and the chemical formulas of the primary ingredient of each are: aspirin ($HC_9H_7O_4$), slaked lime ($Ca(OH)_2$), and the antacids $Mg(OH)_2$ and $CaCO_3$. How can you distinguish between them by means of chemical tests?

2. Bottles labeled 1, 2, 3, and 4 contain solutions of one of four consumer products. The products and the chemical formulas of the primary ingredient of each are: drain cleaner ($NaOH$), vinegar ($HC_2H_3O_2$), milk of magnesia ($Mg(OH)_2$), and rubbing alcohol ($C_3H_7O_7$). Design experiments to distinguish among these possibilities and identify the contents of each bottle.

3. Bottles labeled A, B, and C contain one of three different acids that are primary ingredients in consumer products. Can you design experiments to identify the bottle contents and organize the bottles from strongest to weakest acid? The bottles contain 0.01 M solutions of one of three consumer products. The consumer products and the chemical formulas of the primary ingredient of each are: seltzer water (H_2CO_3), cola beverage (H_3PO_4), and eyewash (H_3BO_3).

PROCEDURE AND OBSERVATIONS

Experiment 9–2 Report

Summarize your findings regarding the question, Can the acid–base characteristics of a compound be predicted from its chemical components? Refer to the experimental data and analysis of results to support your conclusions.

1. Which one of the following compounds do ophthalmologists most likely use for eyewash: H_3BO_3, H_2SO_3, or HNO_3? How do you know? Explain your reasoning, and refer to the collected data from the different teams to support your decision.

2. Solutions of high hydroxide ion concentration are corrosive to all tissue, and if ingested can cause pain and gastric bleeding. A suspension of magnesium hydroxide, called milk of magnesia, can be used internally as an antacid without hazard to tissue. Offer an explanation. Refer to the experimental data to support your explanation.

3. The acid strength of the oxyacids of family 17 in the periodic table are: $HClO > HBrO > HIO$. Explain this fact based on experiment results.

4. Predict which 0.01 M solution has a lower pH and explain why:

    a. H_2S or H_2Se
    b. HNO_3 or HNO_2

5. Oxides of the elements in the periodic table exhibit periodic trends in acid–base properties. Listed below in Table 4 are the oxides of the third period elements. Predict whether each compound:

    a. will be basic or acidic
    b. will be the strongest base or acid

    Refer to the results of this experiment to justify your predictions.

Table 4 Oxides of the Third Period Elements

Na_2O	MgO	Al_2O_3	SiO_2	P_4O_{10}	SO_3	Cl_2O_7

6. Predict the comparative acid strength of the acids below. Refer to the results of this experiment to justify your predictions.

    a. Cl—CH_2CH_2COOH (3-Chloropropanoic acid)
    b. $CH_2CHClCOOH$ (2-Chloropropanoic acid)
    c. CH_3CCl_2COOH (2,2-Dichloropropanoic acid)
    d. I—CH_2CH_2COOH (3-Iodopropanoic acid)

Experiment 10–1

Do Chemicals Duplicate the Effects of a Battery?

INTRODUCTION

It is your good fortune to be a candidate for a position requiring the application of chemistry to the design of diverse battery applications. The application process requires that candidates for the position work in teams to study and compare the impact of a battery and chemicals on aqueous solutions of different substances. You are to write a report summarizing your results and respond to the question, Do chemicals duplicate the effects of a battery?

OBJECTIVES

- Compare products of chemical reactions to those of reactions driven by a battery.
- Design strategies to analyze reactions and identify reacting and nonreacting species.
- Determine reaction products from experimental tests and observations.
- Determine whether batteries reproduce the effects of chemicals.

PROCEDURES

Your study starts in Part I with the analysis of reactions resulting from the addition of chemical reagents to aqueous solutions containing iron(II), iron(III), and iodide ions. You will subsequently study and compare (Part II) the effects of a 9-volt battery and chemical reagents on aqueous solutions containing iron(II) and iodide ions. Your final task in Part III is to self-design tests to determine whether chemical mixtures in aqueous solution are replicating the effects of a battery.

PART I—REAGENT EFFECTS ON IRON(II), IRON(III), AND IODIDE IONS

Refer to the guidelines and information provided below while conducting your study of the effects of chemical reagents on aqueous solutions containing iron(III), iron(II), and iodide ions.

INFORMATION

- Acidified iron(II) and iron(III) ammonium sulfate solutions are prepared in 0.25 M H_2SO_4.
- Hydrogen peroxide is an oxidizing agent (commonly used in bleaches and as an antiseptic).

GUIDELINES

- Use spot plates or small test tubes or vials to conduct your investigations.
- To prepare fresh 1 M H_2O_2, add 1 drop of 3% H_2O_2 to 10 drops of H_2O.

> *Caution:* Do not dump the metal ion solutions down the sink!
> Do not allow the acidified solutions to contact your body!
> If contact occurs, wash the area profusely with water.

Part IA—Iron (III) Ammonium Sulfate and Potassium Thiocyanate

1. Add 5 drops of acidic 0.15 M iron(III) ammonium sulfate ($Fe_2(NH_4)_2(SO_4)_4$) to 5 drops of 0.005 M potassium thiocyanate (KSCN). Describe and record the appearance of the individual solutions and the resulting mixture. Save the mixture.

2. Add 5 drops of acidic 0.15 M iron(II) ammonium sulfate ($Fe(NH_4)_2(SO_4)_2$) to 5 drops of 0.005 M potassium thiocyanate (KSCN). Describe and record the appearance of the individual solutions and the resulting mixture. Save the mixture for later use.

3. Compare the reactants and results from steps 1 and 2. Which ion is needed for the color change occurring in the reaction under investigation (step 1)? How do you know?

4. Assuming the reaction involves the coming together of ions in solution, which ions might have reacted to cause the observed color change (step 1)? Consult the Inorganic Compounds section of the *CRC Handbook of Chemistry and Physics* and record the published characteristics of possible products. Do the published properties verify your choice of reacting ion combinations?

5. Add 1–2 drops of freshly prepared 1 M H_2O_2 to the product mixture of iron(II) ammonium sulfate and potassium thiocyanate (step 2). Record your observations. Did the H_2O_2 react with any ions in solution?

Data Analysis and Implications (Part IA)

1. Thiocyanate (SCN^-) ion is used as an indicator for the iron(III) ion. How? Refer to your observations.

2. Based on your step 5 observations, what chemical species must be present in the product mixture after the addition of H_2O_2? How do you know?

3. Hydrogen peroxide is an oxidizing agent and reacts with reducing agents. Assuming an oxidation-reduction reaction is occurring in step 5, what species is the reducing agent?

4. Write a balanced equation or net equation that represents the reaction between iron(III) ammonium sulfate and potassium thiocyanate and is consistent with the data.

1. To 5 drops of 0.10 M potassium iodide (KI) add 1 drop of starch solution and mix. Then add 5 drops of acidic 0.15 M iron(III) ammonium sulfate $(Fe_2(NH_4)_2(SO_4)_4)$. Describe and record the appearance of the individual solutions and the resulting mixture. Save the mixture.

2. To 5 drops of 0.10 M potassium nitrate (KNO_3) add 1 drop of starch solution and mix. Then add 5 drops of acidic 0.15 M iron(III) ammonium sulfate $(Fe_2(NH_4)_2(SO_4)_4)$. Describe and record the appearance of the individual solutions and the resulting mixture. What do the results tell you about the identity of reacting species needed for the color change in the reaction under investigation (step 1)?

3. To 5 drops of 0.10 M potassium iodide (KI) add 1 drop of starch solution and mix. Then add 5 drops of acidic 0.15 M iron(II) ammonium sulfate $(Fe(NH_4)_2(SO_4)_2)$. Describe and record the appearance of the individual solutions and the resulting mixture. What do the results tell you about the identity of reacting species needed for the color change in the reaction under investigation (step 1)?

4. To 5 drops of 0.10 M potassium iodide (KI) add 1 drop of starch solution and mix. To 5 drops of iodine (I_2) water, add 1 drop of starch solution and mix. Record your observations. What do these observations tell you about the identity of the product species responsible for the color change in the reaction under investigation?

Data Analysis and Implications (Part IB)

5. Starch is used as an indicator for iodine (I_2). How? Refer to your observations.

6. Write a balanced equation or net equation that represents the reaction between iron(III) ammonium sulfate and potassium iodide and is consistent with the data.

Your team is to compare the effect of reagents and a 9-volt battery on aqueous solutions of iron(II) or iodide ion. Are the effects of the reagents the same or different from the effects of the battery? Read the information and follow the guidelines provided below as you conduct your investigation.

INFORMATION

Testing reagents:

- 0.15 M $Fe(NH_4)_2(SO_4)_2$ in 0.25 M H_2SO_4 and 0.005 M KSCN
- 0.02 M $KMnO_4$
- 0.025 M $K_2Cr_2O_7$
- 1 M H_2O_2 or dilute commercial (3%) H_2O_2 10-fold in H_2O before using
- 0.10 M KI and starch solution: mix 5 parts 0.10 M KI to 1 part starch before using

GUIDELINES

- Use a spot plate or small test tubes for your investigation.
- Use a 9-volt battery with copper wires stripped at both ends (or as indicated by the instructor).
- Clean the copper battery wires between trials.

> *Caution:* Chromium and manganese ions are toxic!
> Do not dump any of the metal ions down the sink!
> Do not allow body contact with acidified solutions! If contact occurs, wash profusely with water.

1. Following the pattern shown in Table 1, first add 5 drops of the solution containing iron(II) and SCN^- ions to each of the indicated (third column) wells in a spot plate or 5 small, labeled test tubes. Record the color of this solution mixture in the table (third column, row two). Then add 5 drops of the solution containing iodide (I^-) ions and starch to the indicated (fourth column) wells in a spot plate or 5 small test tubes. Record the color of this solution mixture in the table (fourth column, row two).

Table 1 Battery versus Reagent Effects on Iron(II) and Iodide Ions

	Color ↓	Iron(II) and SCN⁻ Ions	Iodide Ions and Starch
Color →			
$KMnO_4$			
$K_2Cr_2O_7$			
H_2O_2			
Battery			

2. To each of the indicated wells (row three, columns 1–3) in the spot plate or appropriate test tubes, add a drop of the $KMnO_4$ (potassium permanganate) solution and mix. Record in Table 1 (column two, row three) the color of the original $KMnO_4$ solution by itself. Also record (columns 3 and 4, row three) any observed color change for the $KMnO_4$ or for the other solution(s) with which you mixed it. Add a second drop of $KMnO_4$, mix, and repeat your observations. Continue this process for the $K_2Cr_2O_7$ (potassium dichromate) and H_2O_2 (hydrogen peroxide) solutions.

3. Hold a clean copper wire on each terminal of a 9-volt battery. Place the two copper wires into the iron(II) and SCN⁻ solution and observe what happens at each wire. Clean the tips of the copper wires and place the two copper wires into the iodide ions and starch solution. What happens at the wires connected to the + and − end of the battery? Record your observations.

OBSERVATIONS

Data Analysis and Implications (Part II)

7. What happened to the colorless iron(II)/SCN⁻ solution in each case where a change occurred? What happened to the colorless iodide ions/starch solution in each case where a change occurred? In each case where a color change occurred, what species are responsible for the observed changes?

8. What happened to the color of the $KMnO_4$ and $K_2Cr_2O_7$ solutions when added to solutions of iron(II) and iodide ions? If iron(II) and iodide ions are losing electrons, where do the electrons go?

9. The positive terminal of the battery lacks electrons. What reaction occurs at the positive electrode with the iron(II) solution? What reaction occurs at the positive electrode with the iodide ion solution? If iron(II) and iodide ions are losing electrons, where do the electrons go?

10. What differences, if any, can be observed between what happens at the copper wire connected to the positive terminal of the battery and the copper wire connected to the negative terminal when these wires are immersed in the iron(II) ions?

You and your team are to design experimental tests to analyze the reagent mixtures indicated in Table 2. Is a reaction occurring? If so, is any reagent replicating the effect of a battery? For example, is I^- in KI releasing its electrons and forming I_2? Follow the guidelines provided below to conduct your investigation. Record all tests and observations.

GUIDELINES

- Design tests of the systems using any of the solutions provided in Part I and Part II.
- Compare the properties of proposed products with properties listed in the *CRC Handbook* or another chemical reference book.

Table 2 Part III Reagent Mixtures

1. Add 5 drops of 0.10 M KI to 5 drops of 0.10 M $Cu(NO_3)_2$.
2. Add 5 drops of 0.10 M KI to 5 drops of 0.10 M $AgNO_3$.

TESTS AND OBSERVATIONS

Experiment 10–1 Report

Write a report summarizing your answer to the question, Do chemicals duplicate the effect of a battery? Refer to your collected data to support your conclusions.

1. $KMnO_4$, $K_2Cr_2O_7$, and H_2O_2 are oxidizing agents. Common products of reaction are Mn^{2+}, Cr^{3+}, and H_2O. Identify the change in oxidation state for the involved elements.

2. The negative terminal of the battery has an excess of electrons. Since the electrode is negative, it must be giving up an electron to something in solution. Would the observed bubbles be consistent with oxidizing water to O_2 or reducing H^+ to H_2?

3. "Iron-enriched" cereals are often thought to contain iron(III) ions (or perhaps iron(II) ions). However, such cereals actually contain tiny pieces of iron (Fe) metal. The metal reacts with stomach fluid (often described as $\cong 1$ M HCl) to yield metal ions:

$$Fe\,(s) + 6\,HCl\,(aq) \rightarrow FeCl_3\,(aq) + 3\,H_2\,(g)$$

What is the oxidation state of iron and hydrogen and chlorine in the reactants and products? What is oxidized? What is reduced? What is the reducing agent?

4. Nitrogen dioxide (NO_2) is responsible for the brown haze that forms over cities such as Los Angeles, California. This highly toxic gas is produced from many reactions, including the reaction of copper metal and concentrated nitric acid:

$$Cu\,(s) + 4\,H^+\,(aq) + 2\,NO_3^-\,(aq) \rightarrow Cu^{2+}\,(aq) + 2\,NO_2\,(g) + 2\,H_2O\,(l)$$

What is the oxidation state of copper and nitrogen in the reactants and products? What is oxidized? What is reduced? What is the reducing agent?

5. When a pair of copper wires are connected to a 9-V battery and immersed in a solution of I^- ions, one of these wires becomes coated with I_2. Besides wiping the wire off, how else could you remove this coating?

6. The amount of arsenic in a solution can be found by determining the amount oxidized by iodine. For example:

$$As_2O_3\,(aq) + 2\,I_2\,(s) + 2\,H_2O\,(l) \rightarrow As_2O_5\,(aq) + 4\,HI\,(aq)$$

Write the oxidation number above the symbol for any species that changes oxidation state during the reaction. Identify the reducing and oxidizing agents.

Experiment 10-2

Which Metal Is Appropriate for Food and Beverage Storage?

INTRODUCTION

A local company plans to use some of their excess metal to produce cans for food and carbonated beverage storage. They are aware that some metals are highly reactive and dissolve on contact with the acids in food while others do not. They have asked for your help in sorting the group of metals and their metal ions based on their reactivity. The company asks you to recommend metals for can production.

OBJECTIVES

- Determine what kinds of observations can be used to compare the reactivity of a group of metals.
- Determine the reactivity of metals and their metal ions as well as hydrogen and hydrogen ion.
- Write a scientific report to the company summarizing your findings and recommendations.

PROCEDURES

The research plan is to:
1. Become familiar with the tests and observations that can be used to compare metal reactivity (Part I).
2. Design experiments to compare metal and metal ion reactivity (Part II) and metal reactivity in acid (Part III).
3. Recommend metals for can production.

PART I—RANKING A PAIR OF METALS

You are to begin your research by comparing the reactivity of a pair of metals. You need to decide whether each of the metals (Fe and Cu) reacts with the other's metal ion (iron(III) ion and copper(II) ion). When you are done, you will have entered your observations into Table 1 and be able to describe the comparative reactivity (reducing agent strength) of Fe and Cu.

INFORMATION

- The metal ions are provided in 0.10 M aqueous nitrate solutions:

   0.10 M copper (II) nitrate, $Cu(NO_3)_2$, and 0.10 M iron(III) nitrate, $Fe(NO_3)_3$

- Twenty drops $\cong$ 1 mL

GUIDELINES

- Clean all metals with sandpaper or Lion's Tongue cloth to remove any oxide coating.
- Use about 2–5 mL of each metal ion solution in small test tubes, beakers, or wells of a well plate.

> *Caution:* Do not pour metals or metal ion solutions down the drain! Do not allow solutions to contact your skin!

1. Add a clean strip of iron metal to separate solutions of copper(II) nitrate and iron(III) nitrate. Observe what happens, if anything, and record your observations in Table 1.

2. Repeat this procedure with clean strips of copper metal and fresh samples of iron(III) nitrate and copper(II) nitrate. Observe what happens, if anything, and record your observations.

Table 1 Reactivity of Fe and Cu

	Iron	Copper
Iron(III) nitrate		
Copper(II) nitrate		

Data Analysis and Implications (Part I)

1. Which metal is more active (i.e., the better reducing agent): iron or copper? What experimental evidence do you have to support your conclusion?

2. Which metal ion is more active (i.e., the better oxidizing agent): iron(III) ion or copper(II) ion? What experimental evidence do you have to support your conclusion?

3. Write a chemical equation based on what you observed when iron and copper metal strips were immersed in solutions of iron(III) and copper(II) ions. If no observable change occurred, indicate "no reaction."

You are now to design experiments that will allow you to sort a group of metals (zinc, copper, aluminum, iron, and magnesium or a group of metals provided by your supervisor) into a sequence of decreasing (reducing agent strength) activity. Before starting your research, refer to the information provided below and create a table below (Table 2) for recording your tests and observations of the different metal and metal ion combinations.

PART II—CREATING AN ACTIVITY SERIES

INFORMATION

- 0.10 M salt solutions (nitrates) will be provided: zinc nitrate, $Zn(NO_3)_2$; copper(II) nitrate, $Cu(NO_3)_2$; aluminum nitrate, $Al(NO_3)_3$; iron(III) nitrate $Fe(NO_3)_3$; and magnesium nitrate, $Mg(NO_3)_2$.

- Strips of metals will be provided: Zn, Cu, Al, Fe, and Mg.

- Clean all metals with sandpaper or Lion's Tongue cloth to remove any oxide coating.

- Chemists use the greater than symbol, $>$, to describe comparative reactivity (e.g., Na $>$ Ag indicates that Na is more reactive than Ag).

TEAM DATA

Table 2 Creating an Activity Series

Data Analysis and Implications (Part II)

4. Write chemical equations for all the reactions that you observed while sorting the group of metals into a sequence of *decreasing* activity.

5. List the tested metals in order of *decreasing* activity (reducing agent strength). Use the symbol > to record your results. For example, if X metal is more active than Y metal, you would record X > Y. Indicate how the experimental evidence supports the placement of each metal in your activity series.

6. List the tested metal ions in order of *decreasing* activity (oxidizing agent strength). Use the symbol > to record your results.

7. Compare the activity ranking of the metals and metal ions. Is there any pattern? For example, if you are told that metal C is more active than metal B and metal B is more active than metal A, can you predict the ranking of the metal ions A^{2+}, B^{2+}, and C^{2+}?

Now you are to determine the placement of hydrogen ion in the activity series created in Part II. You can meet this goal by immersing a strip of each of the metals used in Part II in a solution of 1.0 M hydrochloric acid, HCl. Before starting your study, create a table below (Table 3) for recording your tests and observations of the different metals in acid. Record which metals react with the acid and which do not.

> *Caution:* Do not allow acid or metal ion solutions to contact your skin! If contact occurs, wash the area profusely with cold water. Notify your instructor of any spills or body contact with solutions.

TEAM DATA

Table 3 Adding Hydrogen Ion to the Activity Series

Optional Extension

If you remove the coatings from the inside of commercial metal cans, will the contents react with the exposed metal? Design an experiment to determine whether the exposed metal reacts with the contents of the can and concentrates in the food or beverage contents as its metal ion.

Data Analysis and Implications (Part III)

8. Based on your observations, where would you place hydrogen, H_2, in your reducing agent reactivity series from Part II? Why?

9. Based on your observations, where would you place hydrogen ion, H^+, in your oxidizing agent reactivity series from Part II? Why?

Experiment 10–2 Report

Provide a scientific report to the company summarizing your findings relative to sorting the metals and their metal ions on the basis of their "activity." State your recommendation regarding selection of metal/s for can production for storing foods and carbonated beverages and include your answer to the following questions:

Is there any useful pattern/s of information in the results? Specifically, is it possible to predict whether a metal and acid (in particular, the ion, H^+) will or will not spontaneously react? For example, can you predict whether Zn will or will not react with the acids in food if you know that $H_2(g) + Zn(s) =$ no reaction?

1. Individuals living near copper mines sometimes earn extra money by suspending iron objects in the wastewaters. Based on your investigation, what is their secret?

2. Cesium metal reacts with salts containing almost every other metal ion, releasing that other metal. Is cesium very active or very inactive? Where would you place it in the activity series that you have constructed?

3. The earliest method of extracting aluminum was to heat aluminum chloride with sodium to give sodium chloride and aluminum. Which metal is more active, aluminum or sodium?

4. Hydrogen, $H_2(g)$, is above silver metal in the activity series of the metals. Will the following reaction, $Ag(s) + H^+(aq) \rightarrow$ occur or not occur? Why or why not?

5. You observe that when copper is added to a solution containing gold ions, the copper dissolves and the gold precipitates. Similarly, almost every other metal will displace gold from solutions of gold salts.

    a. Do these observations mean that gold ion is an unusually good oxidizing agent or a poor oxidizing agent? Explain.
    b. Expensive units of stereo and other electronic equipment use gold-coated parts. Is the choice of gold based strictly on its appearance; that is, do the observations support its selection based on its chemical behavior? Explain.

6. The food and beverage industry coats metal containers used for commercial products.

    a. Search the Web for further details about coatings.
    b. Customers are generally encouraged to not purchase food in dented cans. Does denting harm the protective coating?

Which Metals Provide the Best Voltage?

Your corporation recently received large bequests of four different metals: iron, zinc, copper, and magnesium. These metals and their metal ions can be used to produce electricity. Your corporation wishes to use these metals to produce batteries. A battery requires two metals and their ions. Your task is to investigate these four metals and their metal ions for their potential to provide voltage and recommend a battery design that gives the largest voltage.

INTRODUCTION

- Design a set of experiments for measuring the voltage produced by electrochemical (galvanic) cells constructed from combinations of four metals and solutions of their metal ions.
- Determine whether the magnitude of the cell voltage is predictable from the redox properties of the metals comprising the cell.
- Depict a battery design that gives the largest voltage possible using available materials and some of the four metals and their metal ion solutions.

OBJECTIVES

Before designing your research study of the voltage produced by the different metals and metal ions (Part II), you are to practice setting up and measuring the voltage of an electrochemical cell using the provided materials and equipment (Part I).

PROCEDURES

1. Use the metals and metal ions of Cu and Zn to construct a standard cell following the guidelines provided below. A standard electrochemical cell is depicted in Figure 1. An alternative microscale apparatus may be assembled and is described on page 355.

PART I—
CONNECTING
THE VOLTMETER

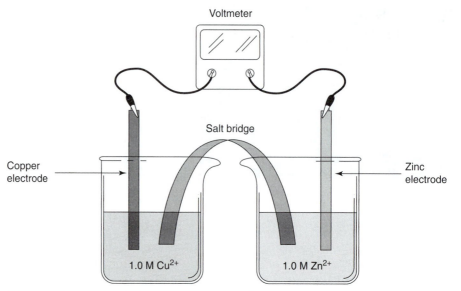

Figure 1 Standard Electrochemical Cell

GUIDELINES

- Metal surfaces must be cleaned with synthetic scrubber, sandpaper, or steel wool prior to use.
- To create a "salt bridge" immerse one end of a *strip* of filter paper in one metal ion solution and the other end into the second metal ion solution. Use a medicine dropper to lay a continuous stream of salt solution (saturated NH_4NO_3 or 1 M $NaNO_3$ or instructor assigned) along the paper between the two half-cells. If filter paper is unavailable, cotton string or folded paper towels soaked in salt solution may be used to connect the half-cells.
- Connect each metal electrode to the voltmeter with a wire that has an alligator clip on one end and a plug that fits the voltmeter terminal on the other.
- Record the average voltage from multiple readings.

> *Caution:* Do not allow metal ion solutions to contact your skin! Wear gloves!
> Do not pour metals or metal ion solutions down the drain!
> Dispose of metals and metal ion solutions in an appropriate waste container!

Standard Electrochemical Cell:

Place about 20 mL of 0.10 M Cu(II) ion solution (0.10 M $Cu(NO_3)_2$) in a small (50 or 100 mL) beaker. Immerse a piece of Cu metal in the Cu(II) ion solution. This is a "half-cell." Create another half-cell using Zn and a nitrate solution of 0.10 M zinc ions. Create a "salt bridge" and connect each metal to one of the terminals of the voltmeter (see guidelines above) and record the *magnitude* (see information on next page) of the voltage.

Microscale apparatus:

Place a piece of filter paper in a glass dish. Put 3 – 4 drops of 0.10 M Cu(II) ion solution (0.10 M Cu(NO$_3$)$_2$) on the filter paper. Put a piece of clean Cu metal in the "puddle." This is a "half-cell." Create another half-cell using Zn and a nitrate solution of 0.10 M zinc ions. Create a "salt bridge" and connect each metal to one of the terminals of the voltmeter (see guidelines above) and record the *magnitude* (see information below) of the voltage.

2. Test the system and record your observations below. For example, what happens to the voltage if you reverse the leads? What happens to the voltage reading if you remove the salt bridge? Which metal ion solution becomes less concentrated during battery operation? Which electrode loses (or gains) mass? Make sure that you read the information provided below and collect enough observations to answer the questions under Data Analysis and Implications.

INFORMATION

- A positive (or more positive) cell voltage indicates that a spontaneous oxidation-reduction reaction is occurring and the (+) lead (also called the cathode) is in contact with the reduction reaction and the (−) lead (the anode) is in contact with the oxidation reaction.
- If a negative (or less positive) voltage reading is observed, then reverse the leads.

PROCEDURE AND OBSERVATIONS

Date: _____ Name: _____ Class: _____

Data Analysis and Implications (Part I)

1. Which metal is the anode (site of oxidation) in this electrochemical cell? Which metal is the cathode and in contact with the reduction reaction? How do you know?

2. In this cell, do the electrons flow from the copper metal to the zinc metal, or vice versa? How do you know?

3. **INFORMATION**

   - Below is the shorthand symbolic diagram for representing a cell

Anode	Salt bridge	Cathode
↓	↓	↓
Metal 1(s)| Metal ion 1(aq)|| Metal ion 2(aq)| Metal 2(s)		
Half-cell		Half-cell
(oxidation)		(reduction)

Note: When we assemble a cell, the part visually to our left won't necessarily be the anode. (Someone on the other side of the lab bench would see the same half-cell on the right). We only say it is the anode if we determine by experiment that oxidation occurs in that half-cell.

Identify (circle) the correct diagram (a or b) for a cell constructed from the metals and metal ions of Cu and Zn based on the information provided above.

a. $Zn(s)| Zn^{2+}(aq)|| Cu^{2+}(aq)| Cu(s)$

b. $Cu(s)| Cu^{2+}(aq)|| Zn^{2+}(aq)| Zn(s)$

4. What would you know if you replaced the zinc strip and zinc ions with another metal and its metal ions and the cell potential changed sign?

Design and perform a set of experiments for determining the voltage produced by electrochemical cells constructed from different combinations of the four metals (Cu, Zn, Fe, and Mg) and their metal ions. Record the potentials you observe, noting carefully which metal electrode is connected to each of the terminals of the voltmeter for each measurement.

INFORMATION

- Metal strips (Cu, Zn, Fe, and Mg) will be available for testing.
- 0.10 M nitrate solutions and solid salts of the metal ions (Cu(II), Zn, Fe(III), and Mg) will be provided.
- Avoid contaminating the solutions! Change the strip used as the salt bridge each time a measurement is taken.

PROCEDURE AND DATA

Data Analysis and Implications (Part II)

5. Provide a record of your team-measured cell potentials in the form of a table in order of *decreasing* cell potentials. Indicate the metal and metal ion combinations and include a balanced equation for each spontaneous reaction.

6. Determine whether there is any useful pattern in your team results relative to cell voltages and the relative activities (strengths as reducing agents) of the metals comprising the electrochemical cell. For example, the relative activities of the investigated metals are: Cu < Fe < Zn < Mg. Do your results indicate that a magnesium anode in combination with a Cu cathode produces a larger or smaller cell voltage than a Fe anode in combination with a Cu cathode?

Experiment 11–1 Report

Write a brief report to the corporation summarizing your findings with regard to the problem under investigation. Depict a battery design indicating the metals and metal ion solutions that will provide the largest voltage possible with the available materials.

1. Use the results of this experiment to predict an oxidation-reduction reaction that involves iron metal that won't be spontaneous: $Fe\,(s) +$ _____ $\rightarrow$ no reaction.

2. The salt bridge was soaked with a saturated solution of NH_4NO_3 or $NaNO_3$. In which direction do the NH_4^+ (or Na^+) ions move when this salt bridge is used to complete the electric circuit in the electrochemical cell in Part I: toward the zinc metal or toward the copper metal? In which direction do the NO_3^- ions move? What is the purpose of the NH_4NO_3-saturated paper in this experiment?

3. No one should be surprised to find that silver electrodes weren't used in this experiment. Assume, however, that the cell potential is -0.45 V when silver is compared with copper with the technique used in this experiment. Where does silver belong in the activity series given in this experiment? (Refer to information for Part II.)

4. Use the cell potentials for the following oxidation-reduction reactions to decide which of these reactions should be spontaneous.

   a. $Al(s) + Cr^{3+}(aq) \rightarrow Al^{3+}(aq) + Cr(s)$ $E = 0.966$ V
   b. $3\,Cr^{2+}(aq) \rightarrow Cr(s) + 2\,Cr^{3+}(aq)$ $E = 0.33$ V
   c. $Fe(s) + Cr^{3+}(aq) \rightarrow Cr(s) + Fe^{3+}(aq)$ $E = -0.70$ V
   d. $3\,H_2(g) + 2\,Cr^{3+}(aq) \rightarrow 2\,Cr(s) + 6\,H^+(aq)$ $E = -0.74$ V

5. The direction of electron flow between metal electrodes is indicated in the cell diagram below. You have a choice of Au, Sn, or Ni for the electrodes. Your chosen metal's ion will be in the electrode compartment. Which metal electrode combination yields maximum voltage given that the reducing agent strength is Ni > Sn > Au? Record your choice by each electrode.

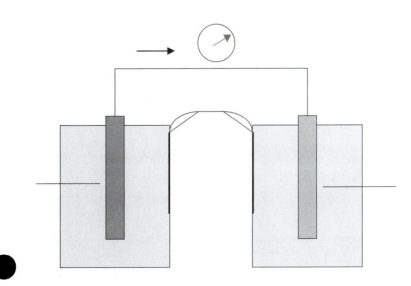

Experiment 11–2

Why Did My Watch Stop Suddenly?

You have recently accepted a consulting position with Timepiece. Your role is to help provide information for customers regarding Timepiece's products. A number of customers have recently inquired as to why a battery-driven watch operates for a period of time and then unexpectedly stops rather than gradually slowing down. Your task is to explore what happens as a battery "runs down" in order to explain why it seems to die so suddenly.

OBJECTIVES

- Investigate the effect of conditions (concentration, temperature, ion charge) on battery voltage.
- Create a graphical plot of your data showing the effect of conditions on voltage.
- Develop an equation from data demonstrating the effect of conditions on voltage.
- Write a scientific report explaining why a battery-driven watch stops suddenly rather than gradually slowing down.

PROCEDURES

Teams are to study the effect of concentration (Part I), temperature (Part II), and ion charge (Part III) on cell voltage. In order to collect, compare, and validate data quickly for multiple conditions, your Timepiece supervisor asks that different teams study different battery conditions and share voltage results. The teams are to collect voltage data using different concentrations of HCl (Group A or B, Table 1 as assigned).

Team Materials

1 pH meter capable of measuring to 1 mV.

Two standard pH electrodes *or* two graphite electrodes

4 100-mL beakers

50 mL of HCl and Cu(II) ion solutions at 1.0 M and 0.10 M, respectively

Pipet or buret for delivery of ± 0.05 mL of solution

Team Materials (continued)

50 mL volumetric flasks for dilution of solutions

6 strips of chromatography paper or another form of absorbent paper, 10 cm × 2 cm

5 mL of a saturated solution of KCl in water

A medicine dropper to add the saturated solution of KCl to the chromatography paper

A large crystallizing dish, trough, or beaker that can be used as a temperature bath

Table 1 Team Assignments: HCl Concentrations

Group A: 1.0 M HCl	Group B: 0.10 M HCl
vs. 0.10 M, 0.01 M, and 0.001 M HCl	vs. 0.01 M, 0.001 M, and 0.0001 M HCl

A source of warm water (about 40°C) *or* ice

PART I—THE EFFECT OF CONCENTRATION (HYDROGEN ION)

What will happen to the cell voltage as the concentrations of reactants decrease? The teams are to collect voltage data using a cell containing solutions of HCl of different concentrations (Group A or B as assigned).

1. What do you expect to happen to the voltage of the cell as you decrease the concentration of ions in one cell? Record your hypothesis.

Hypothesis

2. Deliver 5.0 mL of the stock solution (1.0 M or 0.10 M HCl) into a *clean* 50-mL volumetric flask. Fill the flask with water to the 50-mL calibration line to give 50 mL of (0.10 M or 0.01 M HCl) solution. Prepare 50 mL of 0.01 M or 0.001 M HCl solutions in a similar manner. *Save these solutions for use in Part II.*

> *Caution:* HCl can cause eye damage! Safety goggles must be worn at all times!
> HCl can cause skin damage! If contact occurs, wash the area with water.

3. Figure 1 depicts a cell connected to a pH meter for voltage measurement. For the purpose of sharing different team results, all teams need to follow the guidelines below for assembling a cell.

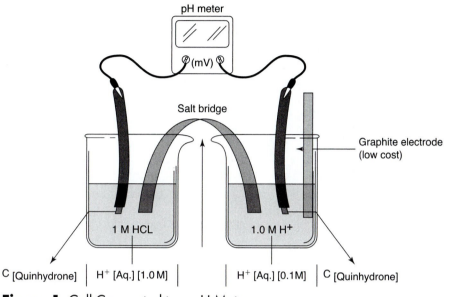

Figure 1 Cell Connected to a pH Meter

GUIDELINES

- Use commercial pH electrodes connected to a pH meter or use carbon (graphite) electrodes.
- If graphite electrodes are used, add a "pinch" of quinhydrone to each acid solution to stabilize the voltage.
- Connect each metal electrode to the voltmeter with a wire that has an alligator clip on one end and a plug that fits the voltmeter terminal on the other.
- To create a salt bridge, use a medicine dropper to soak a piece of absorbent paper with a saturated solution of KCl. Immerse one end of each paper strip into each solution.
- To avoid contamination of solutions, use a different piece of filter paper with saturated KCl to create a salt bridge for each pair of solutions.

4. All teams should record testing procedures and data on concentration and cell voltage (mV) in Table 2. When you are done, compare your results with those of any other team testing the same group of solutions. If the potential (mV) data is comparable (or almost comparable), there is no need to take additional readings; otherwise, repeat your test procedure until the readings are consistent.

5. When you are done, compile the results of the different teams using a computer or Table 3. Do the results support your hypothesis?

TEAM DATA

Table 2 Concentration and Cell Voltage at Room Temperature

Cell ____ M versus ____ M HCl	Potential (mV)	Concentration Difference[*]	Concentration Ratio[**]	Natural Log of Conc. Ratio

[*]Concentration difference: e.g., Group A = 1.0 M − 0.10 M (or Group B = 0.10 M − 0.01 M)
[**]Concentration ratio: e.g., Group A = 1.0 M/0.10 M (or Group B = 0.10 M/0.01 M)

Table 3 Concentration and Cell Voltage at Room Temperature

Cell _____ M vs. _____ M HCl	Potential (mV)	Concentration Difference	Concentration Ratio	Natural Log of Conc. Ratio

Data Analysis and Implications (Part I)

1. Examine the compiled team data, and describe any pattern between concentration and cell voltage. For example, as concentration increases, does voltage increase or decrease? Indicate whether the results support your hypothesis.

2. Use the compiled team data to create a graph (attach) for each of the bulletted items, for interpreting results:

    -   Potential (mV) vs. the difference between the concentrations of the tested solutions
    -   Potential (mV) vs. the ratio of the concentrations of the tested solutions
    -   Potential (mV) vs. the natural logarithm of the ratio of the concentrations of the tested solutions

3. Do any results offer a useful relationship and possible explanation for the problem under investigation: Why did my watch stop suddenly? Why or why not? Explain.

In Part I you measured the potential difference between cell electrodes produced by HCl solution pairs at room temperature. What will happen, if anything, to the voltage of the cell if you alter the cell temperature?

Your team is now to measure the potential difference (mV) using the same HCl solution concentration pairs used in Part I at 10°C or 40°C (or another temperature as assigned by your supervisor).

1. What do you expect to happen, if anything, to the voltage of the cell as you alter the temperature? Record your hypothesis.

   Hypothesis

2. Read the guidelines below before proceeding with experimentation. Record your team data in Table 4. When you are done, compare your results with those of any other team testing the same group of solutions at the same temperature. If the potential (mV) data are comparable there is no need to take additional readings; otherwise, repeat the test procedure until the readings are consistent. When you are done, compile the results of the different teams using a computer or Table 5.

GUIDELINES

- If you work at 10°C, place the beakers in a large dish containing a mixture of ice and water.
- If you work at 40°C, place the beakers in a temperature bath or large dish with warm water.
- Wait until a thermometer indicates that the solutions are at the same temperature as the water bath before making any measurements of potential.

> *Caution:* HCl can cause eye and skin damage!
> If contact occurs, wash the area with water.

TEAM PROCEDURE AND DATA

Table 4 Concentration and Cell Voltage at _____°C

Cell _____ M vs. _____ M HCl	Potential (mV)	Concentration Difference	Concentration Ratio	Natural Log of Concentration Ratio

COMPILED TEAM DATA

Table 5 Concentration and Cell Voltage versus Temperature

Temperature (°C)	Potential (mV)	Concentration Ratio	Natural Log of Concentration Ratio

Data Analysis and Implications (Part II)

4. Create a graph (attach) of the compiled team results of potential measurements versus the natural logarithm of the ratio of the solution pair concentrations at the different temperatures (Table 5). Answer the following questions:

   a. What happens to the average value of the slope of the straight line in these graphs as the temperature is increased from 10°C to about 40°C? Record your calculations.

   b. Do the results support your hypothesis? Why or why not?

5. Based on the results, do temperature differences affect potential? For example, will the same battery produce a different voltage outdoors in Alaska and in Africa? Why or why not?

What effect does ion charge have on cell voltage? Your optional task is to collect voltage data using a cell containing solution pairs of Cu(II) ions at the same concentrations (Group A or Group B as assigned) previously studied for hydrogen ions.

Group A: 1.0 M Cu(II)	Group B: 0.10 M Cu(II)
vs. 0.10 M, 0.01 M, and 0.001 M Cu(II)	vs. 0.01 M, 0.001 M, and 0.0001 M Cu(II)

1. What do you expect to happen to the voltage of the cell as you decrease the concentration of Cu(II) ions? Do you expect any voltage changes to be similar to or different from those observed with decreases in concentration of hydrogen ions (Part I)? Record your hypothesis.

   Hypothesis

2. Read the guidelines below, and then proceed to collect voltage data using a cell containing solution pairs of Cu(II) ions at the same concentrations you previously studied for hydrogen ions in Part I.

GUIDELINES

- Use two copper metal electrodes instead of commercial pH or carbon (graphite) electrodes.
- Clean the Cu electrodes prior to use with sandpaper, or immerse the strips in 1M HCl for about two minutes. Rinse the Cu strips with deionized water, and then blot them dry.
- Record the potential difference between the electrodes in units of millivolts (mV) for each solution pair at room temperature.

> *Caution:* Avoid ingesting Cu(II) ions! Cu(II) ion can be toxic in large quantities!
> Do not pour Cu(II) ion solutions down the drain!
> Dispose of Cu(II) ion solutions in an appropriate waste bottle!

Date: _____ Name: _____ Class: _____

Data Analysis and Implications (Part III)

6. Compare the results for Cu(II) ion (Part III) with those obtained for hydrogen ion (Part I). Are the voltage changes observed with changes in concentration of Cu(II) ion similar to or different from those observed with decreases in concentration of hydrogen ion? Do the results support your hypothesis? Why or why not?

7. Graph the potential measurements for the Cu(II) ion on a graph similar to the one used for hydrogen ion (attach your graph). How does the average value of the slope of the line obtained for the Cu(II) ion compare with the data obtained in Part I for the hydrogen ion?

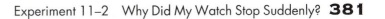

Experiment 11–2 Report

Provide a brief scientific report and advice relative to the common customer complaint: "I am disappointed with Timepiece because I was inconvenienced when my watch unexpectedly ran down." Summarize your findings with regard to the problem, Why does a battery-driven watch stop suddenly? Make sure you refer to experimental observations, data, and graphs to support your conclusions.

1. This question refers to the electrochemical cell $Cu/Cu^{2+} // Pt^{2+}/Pt$ where the cation solutions are 1.0 M and the potential measures 775 mV. Determine whether the potential will increase or decrease or remain the same if the cation concentrations are:

$$2.0\,M\,Cu^{2+}/2.0\,M\,Pt^{2+} \quad \text{or} \quad 1.0\,M\,Cu^{2+}/2.0\,M\,Pt^{2+}$$

2. The voltage of a battery with 1 M solutions of both Zn^{2+} and Cu^{2+} ions at $25°C$ based on the redox reaction $Zn(s) + Cu^{2+}(aq) \rightarrow Zn^{2+}(aq) + Cu(s)$ is 1.10 V. As the redox reaction proceeds, what happens to the concentration of Cu^{2+} ion? What happens to the Zn^{2+} ion concentration? Can such concentration changes account for a run-down battery?

3. Do the experimental data indicate that the slope of the line resulting from a plot of potential vs. natural logarithm of the ratio of the cell concentrations will be different or the same for all ions of the same charge (i.e., not just for the ions you investigated)?

4. Compare the average value of the slope of the line at the three temperatures with the value of RT/F at each temperature. Are they the same or different? Obtain the value of the ideal gas constant (R) in units of joules per mole-kelvin and the value of the Faraday constant (F) in units of coulomb per mole. Use the fact that one volt is one joule per coulomb to calculate the units of R/F. Multiply the ratio R/F by the absolute temperatures in kelvin.

5. Predict the effect on the potential for the following cell of an *increase* in the pH of the solution. Predict the effect of an *increase* in the H_2O_2 concentration.

$$2\,MnO_4^-(aq) + 5\,H_2O_2(aq) + 6\,H^+(aq) \rightarrow 2\,Mn^{2+}(aq)$$
$$+ 8\,H_2O(l) + 5\,O_2(g)$$

6. Derive an equation that relates the difference between the potential of a pair of solutions (DV), the ideal gas constant (R), Faraday's constant (F), the absolute temperature at which data are collected (T), and the natural log (ln) of the ratio of the concentrations of the two solutions ($[X_1]$ and $[X_2]$).

   a. What would be the potential difference if we studied two solutions that had the same concentration? Show that the answer derived from common sense is the same as the one derived mathematically from your equation.

   b. If the concentration of one solution is zero, what does the equation suggest the potential will be? Is this reasonable?

Experiment 11-3

Why Is a "Harley" Chrome-Plated?

INTRODUCTION

There is something about a Harley Davidson that makes it different from all of the other motorcycles on the road. It might be the sound of that twin-cylinder engine when you fire it up or the beautiful paint job that only Harley seems to achieve. On the other hand, the chrome plate that makes the motorcycle shine also seems different. Whatever it is, these motorcycles are unique.

In this investigation, you will probe the mechanism of the rusting of iron to understand the effect of plating iron with metals such as chromium. Ultimately, you will be asked to decide: Why is a Harley chrome-plated? Is there another reason for chrome-plating some of the metal parts of these motorcycles besides making the motorcycle look "cool"?

OBJECTIVES

- Examine and understand the mechanism of rusting of iron.
- Determine whether metal interactions may cause iron metal to corrode less or more rapidly than normal.
- Determine whether metal corrosion rate correlates with metal reduction potential.
- Determine whether a Harley is chrome-plated for protective or decorative purposes or both.

PROCEDURES

Your goal is to understand the mechanism of rusting of iron and determine if metals such as chromium alter the rate of iron rusting. In Part I you will study the mechanism of rusting. In Part II you will determine whether different metals alter the rate of iron rusting.

During rusting does iron convert to iron(II) and/or iron(III) ions? In Part IA you will become familiar with tests that allow you to distinguish between and identify iron(II) and iron(III) ions. In Part IB you will perform tests to determine the identity of iron products formed during the rusting of iron.

Part IA—Qualitative Tests for Iron(II) and Iron(III) Ions

1. Obtain about 20 mL of 0.010 M $Fe(NH_4)_2(SO_4)_2$ (Fe(II) solution) and 20 mL of 0.010 M $Fe_2(NH_4)_2(SO_4)_4$ (Fe(III) solution) in clean, dry, labeled beakers. Also obtain about 20 mL of 2 M H_2SO_4, 5 mL of 1.0 M KSCN, and 5 mL of 1.0 M $K_3Fe(CN)_6$ in clean, dry, labeled beakers.

2. Place about 5 mL of water, 5 mL of Fe(II) solution, and 5 mL of Fe(III) solution in separate clean, dry, labeled small test tubes or well plates. Record the appearance of each solution in Table 1 below. Add 1 mL of 2 M H_2SO_4 to each and record their appearance. Finally, add 1 drop of 1.0 M KSCN to each solution and record your observations.

3. Repeat the above procedure (step 2) using 1 drop of 1.0 M $K_3Fe(CN)_6$ instead of KSCN and record your observations in Table 1.

Table 1 Observations of Fe(II) and Fe(III) Ions

	Water	Fe(II) Ion	Fe(III) Ion
H_2SO_4			
SCN^-			
$Fe(CN)_6^{3-}$			

4. Before proceeding to Part IB, answer the following question based on your observations. How can SCN^- and $Fe(CN)_6^{3-}$ be used to identify the presence of Fe(II) and/or Fe(III) ions?

1. Prepare 10 iron nails (or iron strips) to study the rusting of iron. Make sure that the nails *are not* galvanized and contain no rust! Clean the surfaces of the iron nails or strips with sandpaper or a scouring pad and/or by immersing them in a 2 M HCl solution in a fume hood for a few minutes. Rinse the nails and dry them.

> *Caution:* HCl can cause eye damage and burns if it contacts your skin! Wear safety goggles!
> Wash your skin immediately with soap and water if contact occurs!

Use four clean nails for these qualitative observations and save the rest for the Optional Extension (if assigned) and Part II.

2. Place several layers of white absorbent paper in four large petri dishes or recyclable plastic plates. Saturate the paper with 0.02 M NaCl solution and then drain out the excess liquid. Place a clean iron nail onto the surface of the absorbent paper in each of these dishes and push down on these nails to create a depression in the absorbent paper. Make sure that the entire length of the nail makes contact with the surface of the absorbent paper.

3. Remove one of the nails and add 7–10 drops of SCN^- ions (1.0 M KSCN) along the depression in the absorbent paper. Use forceps to replace the nail in the depression. Repeat this process by adding drops of phenolphthalein beneath the nail in the second dish and drops of potassium ferricyanide $[K_3Fe(CN)_6]$ beneath the nail in the third dish. The fourth dish serves as a control, and therefore nothing should be added so the iron nail is only in contact with the absorbent paper moistened with NaCl solution.

4. After about 5 minutes, remove the nails—one at a time—and record your observations after reading the information provided below. Repeat this process every 10 minutes for 30–40 minutes, replacing the nail in the depression.

INFORMATION

- A water-soluble acidic product will form H^+ ions in aqueous solution and gives a colorless solution when phenolphthalein is added.
- A water-soluble base product will release OH^- ions in aqueous solution and gives a pink solution when phenolphthalein is added.

Optional Extension   Prepare a test dish combining phenolphthalein and the ferricyanide and a clean nail. What do you observe?

Data Analysis and Implications (Part I)

1. How can SCN^- and $Fe(CN)_6{}^{3-}$ be used to identify the presence of Fe(II) and/or Fe(III) ions?

2. What effect, if any, does NaCl solution have on the rate at which iron rusts?

3. Does the acidity (amount of H^+ ions) or alkalinity (amount of OH^- ions) of the water solution alter during the rusting process? How do you know?

4. What form of iron ions are produced during the rusting process? How do you know?

5. Is iron oxidized or reduced during the rusting process? Complete the equation to show the product species produced during rusting, and identify the oxidizing and reducing agents.

$$Fe\,(s) + O_2\,(g) + HOH\,(l) \rightarrow$$

Do metal interactions alter the rate of iron rusting? Four common (and somewhat inexpensive) metals (Mg, Zn, Sn, and Cu) will be provided for your study in addition to iron. The metals have different tendencies to lose electrons and convert to metal ions (see information below).

INFORMATION

The standard oxidation potential for the half-reactions of the metals to be tested is:

Half-reaction	Potential (V)
$Mg\,(s) \rightarrow Mg^{2+}(aq) + 2\,e^-$	2.372
$Zn\,(s) \rightarrow Zn^{2+}(aq) + 2\,e^-$	0.7618
$Fe\,(s) \rightarrow Fe^{2+}(aq) + 2\,e^-$	0.447
$Sn\,(s) \rightarrow Sn^{2+}(aq) + 2\,e^-$	0.1375
$Fe\,(s) \rightarrow Fe^{3+}(aq) + 3\,e^-$	0.037
$Cu\,(s) \rightarrow Cu^{2+}(aq) + 2\,e^-$	−0.3419

1. Before starting your research, record a hypothesis with regard to the effect of metal interactions on the rate of iron rusting.

Hypothesis

2. Based on your hypothesis, record a prediction regarding the comparative rate of rusting of iron in contact with magnesium to that of iron in contact with copper.

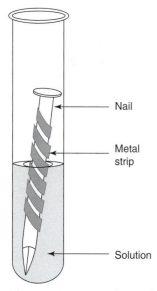

Figure 1 Test Tube with Metal Strip Around Iron Nail

3. Clean and dry the different metal strips. Wrap the clean and dry metal strips around four of the five nails, as shown in Figure 1 and insert the nails into labeled (e.g., iron, iron + magnesium, ...) test tubes set in a rack. The test tube with the iron nail alone is the control. Pour approximately 10 mL of 0.2 M NaCl solution into each of the five test tubes.

> *Caution:* Take care preparing metal strips—the edges may be sharp enough to cut your skin!

4. Record your observations at about half-hour intervals for several hours. For example, are there any indications of rusting? Do you observe any color characteristic of iron ions? Is the alkalinity or acidity of the water changing? While you are waiting, design and conduct some tests to help confirm whether rusting is occurring.

OBSERVATIONS AND TESTS

Data Analysis and Implications (Part II)

6. Do the results support your hypothesis regarding comparative rate of rusting of iron in contact with magnesium to that of iron in contact with copper? Why or why not?

7. Is there evidence of rusting for the nails wrapped with magnesium or zinc or tin or copper? Indicate whether each metal has a positive, negative, or no effect on the *rate* at which iron rusts.

8. Does the rate of iron rusting relate to the oxidation potential of the metal in contact with iron? If so, how? Explain.

9. Chrome plating of iron (or steel) is done in two steps. Generally, first a layer of nickel is plated onto the iron and then chromium metal is plated on top of the nickel. The oxidation potential for the nickel half-reaction and the chromium half-reaction are:

$$Ni(s) \rightarrow Ni^{2+}(aq) + 2\,e^- \qquad 0.26\,V$$
$$Cr(s) \rightarrow Cr^{3+}(aq) + 3\,e^- \qquad 0.74\,V$$

Based solely on this information and the results of this study, will chrome-plating protect iron from rusting? Will the layer of nickel that is initially plated onto the iron protect the iron from rusting? Why or why not?

Experiment 11–3 Report

Write a brief report summarizing your answer to this question: Why is a Harley chrome-plated? Refer to your *observations and experimental data* to support your conclusions. If you feel the answer to the problem is not conclusive or is unclear, suggest other investigations one could pursue.

1. A garbage can is produced by immersing iron metal in molten tin. Another garbage can is produced by immersing tin metal in molten zinc. Which can is less likely to corrode? Why?

2. Is the rusting of iron pH-dependent?

3. The Statue of Liberty consists of copper plates supported on an iron skeleton. The statue was recently repaired. It was only necessary to replace the iron bars or copper plates that had contact. What happened? What was replaced—the copper plates or the iron skeleton? Why?

4. *Cathodic protection* occurs when iron is transformed from the anode to the cathode of an electrochemical cell by connecting iron to a sacrificial anode. Is the sacrificial anode metal a more or less active metal than iron? Give an example of a sacrificial anode metal.

5. The material most commonly used for filling decaying teeth is a composition known as dental amalgam, a solid solution of tin and silver in mercury. If you bite a piece of aluminum foil from a gum or candy wrapper, your tooth nerves will feel a jolt of pain due to an electrochemical reaction. The amalgam consists of a mixture of three solid solutions having different standard reduction potentials: Hg_2^{2+}/Ag_2Hg_3, 0.85 V; Sn^{2+}/Ag_3Sn, -0.05 V; Sn^{2+}/Sn_8Hg_3, -0.13 V. The standard reduction potential of Al is $+1.66$ V. Why do your nerves feel a jolt of pain?

Experiment 12–1

How Long Will It Take?

An often-asked question is: How long will it take? For example, this question is posed when someone wants to know the time required to reach a destination by car. The answer is dependent on many factors including distance, car speed, route chosen, and time of day, among others. A similar question is posed when chemists ask: How long will it take for reactants to combine and form products? The answer describes the rate of the reaction and is dependent on its mechanism and factors such as catalysis, concentration, ionic medium, surface area, temperature, pressure, and solvent. A kinetic study experimentally explores the factors affecting reaction rate.

You and your teammates are working in a research setting where you are required to understand the kinetics of the reaction of magnesium with hydrochloric acid:

$$Mg(s) + 2HCl(aq) \rightarrow H_2(g) + MgCl_2(aq)$$

Your team needs to monitor the rate of reaction of magnesium metal and determine the rate law for the reaction. You also need to determine the impact of temperature on rate.

- Design experiments to measure the rate of disappearance of magnesium metal as a function of time.
- Determine the effect of initial concentration of HCl and Mg on reaction rate.
- Determine the form of the rate law for the reaction of magnesium with hydrochloric acid.
- Summarize your findings in a written report.

The research plan is to:
1. Design experiments to determine reaction rate where different terms use different initial concentrations of acid or Mg (Part I) and share results.
2. Design team experiments and share results regarding the effect of temperature on rate (Part II)
3. Determine the rate law for the reaction of magnesium with hydrochloric acid.

PART I— CONCENTRATION STUDIES AND THE RATE EXPRESSION

In the reaction under study, you can assume that the rate depends on the concentration of acid and on the amount of magnesium metal. Thus, rate = $k[Mg]^x[HCl]^y$.

The rate of reaction will be determined by the initial rate method. The idea is to determine the instantaneous rate of reaction before the initial concentrations of reactants have changed significantly. You will conduct several experiments using different initial reactant concentrations, and measure the reaction rate. You will determine the reaction rate by measuring the time it takes for a known amount of Mg to disappear ($-\Delta[Mg]/\Delta t$). By substituting the values of initial reactant concentrations in the rate law of each trial, the value of the rate constant, k, can be calculated.

PART IA—ACID CONCENTRATION EFFECTS

Teams will examine the impact of acid concentration on reaction rate, with different teams using different initial acid concentrations and sharing results. Conduct rate studies at a minimum of three different acid concentrations or as assigned.

Table 1 Team Assignments: Initial HCl Concentration

Group 1	0.50 M	1.0 M	1.5 M
Group 2	0.80 M	1.2 M	2. 0 M

Note that your research results are only valid if all variables affecting the rate of reaction other than acid concentration are controlled. For example, the reaction of Mg with hydrochloric acid is exothermic and since temperature affects reaction rate, the temperature of the reaction mixture should be controlled. This can be reasonably accomplished by placing the reaction vessel in a water bath at room temperature.

1. Place a 5-cm-long strip of Mg metal into a 100-mL beaker containing 20 mL of 1.0 M HCl and observe the reaction. Based on your observations and knowledge, what do you expect to observe with regard to the time it takes for magnesium to disappear as you change the concentration of the acid? Record your hypothesis.

> *Caution:* Do not allow acid to contact your skin!
> If contact occurs wash the area profusely with cold water.
> Notify your instructor of any spills or body contact with HCl.

Hypothesis

2. Obtain 20 mL of hydrochloric acid at concentrations between 0.50 and 2.0 M as assigned and 5-cm-long pieces of magnesium of known mass. Record the mass. If the length of all the pieces of magnesium is the same, weighing one piece will give you a relatively good average mass. Alternatively, measure the mass of several 5-cm-long pieces of Mg ribbon and calculate and record the average mass.

3. All teams are to use a 100-mL beaker set in a water bath, 20 mL of hydrochloric acid, and a 5-cm piece of Mg ribbon. Record the average initial mass of Mg and concentration of hydrochloric acid in Table 2. For sharing results of different teams, all teams need to follow the guidelines below.

GUIDELINES

- Clean Mg with sandpaper or synthetic scrubber before measuring and recording the mass of a strip.
- Use a stop or digital watch with increments of seconds to record the time it takes for the Mg to disappear; record the time in seconds (s) in Table 1.
- Repeat each study until two results are consistent and an average rate determined.
- All reaction mixtures should be stirred cautiously but continuously with a stirring rod or magnetic stirrer. A magnetic stirrer should *not* be used if it is heat-generating.

> *Caution:* Hydrogen gas is explosive in the presence of air and an open flame!
> No open flames are allowed while conducting this experiment.

TEAM DATA

INFORMATION

- Rate of reaction based on time for disappearance of Mg = $(-\Delta[Mg]/\Delta t)$
- Δt = time (seconds) required for disappearance of Mg
- [Mg] = [moles/liter Mg] calculated from initial average mol Mg in 20 mL HCl

Table 2 Rate of Reaction of Mg versus Acid Concentration

[HCl] (M)			
Time trial 1 (s)			
Time trial 2 (s)			
Average time (s)			
Average rate ($-\Delta[Mg]/\Delta t$)			

Average initial mass Mg: _____ g in 20 mL

Average initial mol Mg: _____ mol in 20 mL

Average initial [Mg]: _____ mol/L

Temperature _____ °C

3. Collect and compile the different team results using a computer or Table 3.

COMPILED TEAM DATA

Table 3 Rate of Reaction of Mg versus [HCl]

Average Rate ($-\Delta[Mg]/\Delta t$)	0.5 M HCl	0.8 M HCl	1.0 M HCl	1.2 M HCl	1.5 M HCl	2.0 M HCl

Data Analysis and Implications (Part IA)

1. Examine the different team results for any pattern with regard to rate of reaction of Mg with increasing concentration of HCl. What, if anything, do the different team results indicate?

2. Do the different team results support your hypothesis? Why or why not?

In Part IA you measured the effect of different initial concentrations of acid concentration on reaction rate. You will now design and conduct an experiment to measure the effect of different initial amounts of magnesium on the rate of reaction.

1. What do you expect will happen to the reaction rate as you increase the amount of magnesium? Record your hypothesis.

   Hypothesis

2. Follow the guidelines below to design and conduct your experiment. Record your team data in Table 4. Make sure to record team-testing conditions, including the variable being studied and the factors held constant. When you are done, compile the results of the different teams using a computer or collect results in Table 5.

GUIDELINES

- Conduct a minimum of three rate studies with three different initial amounts of Mg, where each study is repeated until two results are consistent and an average rate determined.
- Use clean pieces of Mg weighed to ± 0.1 g and with a minimum mass of 0.015 g to a maximum mass of 0.070 g.
- Use 20 mL of 1.0 M (or a concentration indicated by your instructor) hydrochloric acid.
- Record the rate of reaction in seconds based on time for disappearance of Mg.

> *Caution:* Hydrogen gas is explosive in the presence of air and an open flame! No open flames are allowed during this study.

PROCEDURE AND TEAM DATA

Table 4 Rate of Reaction of Mg versus Amount of Mg

	Trial 1	Trial 2	Trial 3
Mg (g)/20 mL			
Time trial 1, (s)			
Time trial 2, (s)			
Average time, (s)			
Average rate* $(-\Delta[Mg]/\Delta t)$			

Temperature_____ °C   [HCl] = _____ M

*Average rate $(-\Delta[Mg]/\Delta t)$ calculations:

	Trial 1	Trial 2	Trial 3
Initial mol Mg/20 mL (mol)			
Initial [Mg] (M)			
Time (s) for disappearance of Mg			
Average rate $(-\Delta[Mg]/\Delta t)$ $(-M/sec)$			

Table 5 Rate of Reaction of Mg versus Concentration of Mg

Team	Initial [Mg]	Time (s)	Average Rate $(-\Delta[Mg]/\Delta t)$

Data Analysis and Implications (Part IB)

3. Examine your team results and the results of other teams for any pattern with regard to rate of disappearance of Mg versus increasing amounts of the metal. What do the results indicate?

4. Do the compiled team data support your hypothesis? Why or why not?

5. Based on the compiled data of the different teams (Tables 3 and 5), determine the values of x, y, and k in the rate expression, rate $= k \, [Mg]^x[HCl]^y$. Record any calculations. If necessary, use the back of this page.

Optional Extension

Integrated forms of rate laws provide another way of determining the order of reaction graphically. Use the compiled data and graphical methods to determine the order of the reaction. For example, plot rate vs. [Mg] (or [HCl]), time vs. natural log [Mg] (or natural log [HCl]), ln (1/time) vs. ln [Mg] (or ln [HCl]). Are there any significant differences between the instantaneous rate law method results and those obtained by the graphical depiction of the integrated form of the rate law?

Design and conduct an experiment to determine the effect of temperature on the rate constant k in the expression, rate $= k[Mg]^x[HCl]^y$.

1. What do you expect to happen, if anything, to the value of the rate constant as you alter the temperature above or below room temperature? Record your hypothesis.

    Hypothesis

2. Follow the guidelines below for designing and conducting your experiment. For the purpose of sharing the results of different teams, all teams are also asked to record the conditions of experimentation (i.e., factors held constant and the experimental variable). When you are done, collect and compile the different team results at different temperatures using a computer or Table 6.

    > *Caution:* Hydrogen gas is explosive in the presence of air and an open flame! No open flames are allowed during this study.
    > HCl can cause eye and skin damage! If contact occurs, wash the area with water.

GUIDELINES

- Repeat each rate study until two rate results are consistent.
- Record the rate of reaction in seconds based on time for disappearance of Mg.
- Use 20 mL of 1.0 M (or an instructor-indicated concentration) HCl.
- Conduct studies at one assigned temperature (10°C, 30°C, or 40°C).
  - At 10°C, place the beakers in a large dish containing a mixture of ice and water.
  - At 30°C or 40°C, place the beakers in a temperature bath or large dish with warm water.
- Do not add Mg until a thermometer indicates the acid is at the desired temperature.

Table 6 Rate of Reaction of Mg versus Temperature

Team	Average Rate $(-\Delta[Mg]/\Delta t)$	10°C	30°C	40°C

Data Analysis and Implications (Part II)

5. Determine the value of k in the rate expression, rate $= k \, [Mg]^x[HCl]^y$ at the team-assigned temperature. Do the results support or refute your hypothesis? Why or why not?

6. Based on the compiled team data, what happens to the rate values of k as the temperature is increased every ten degrees (from 10°C to 40°C)? Is there any pattern? Explain.

Experiment 12–1 Report

How long does it take for reactants to combine and form products in the reaction between Mg and HCl? Provide a brief report describing experiment results, including the effects of initial reactant concentrations and temperature on reaction rate and the rate constant k. What is the rate law for the reaction? How do you know?

1. A team independently repeats this experiment and makes several mistakes. Consider what effect each of the errors would *independently* have on the determined rate of reaction.

   a. The temperature of the reaction mixture in Part IA is not controlled—that is, the beaker is not placed in a water bath.
   b. The concentration of HCl in Part IA is actually lower than 1.0 M.
   c. The reaction mixture is *not* stirred.

2. Physiological processes in humans (and plants and animals) are chemical in nature and thus are affected by temperature in much the same way as the rates of chemical reactions. For example, inflamed athletic injuries are treated by applying an ice pack as soon as possible. After about one to two days, heat packs are sometimes applied. Why? What must be true about the effect of temperature on the rate of biochemical reactions causing inflammation? What must be true about the effect of temperature on the rate of biochemical reactions causing healing?

3. The field of "human kinetics" has similarities to "chemical kinetics." For example, the U.S. Olympic Swim Team Association uses principals of human kinetics during training. Search the Web for information on human kinetics, and report on connections between human and chemical kinetics.

4. The reaction, $A + 2B \rightarrow C + OH^-$, yields the following data:

   At zero time: [A] = 1. M, [B] = 0.10 M, [C] = 0 M, pH = 3.0.
   At 5 minutes: pH = 7.0.

   What is the rate of loss of B in $mol^{-1} L^{-1} min^{-1}$?

5. The table below gives experimental data for kinetic studies of the following reaction:

   $$3A + 2B \rightarrow 2C$$

   What is the rate expression for the reaction?

Trial	Initial [A]	Initial [B]	Rate (mol l^{-1} sec^{-1})
1	0.01	0.02	0.000054
2	0.03	0.02	0.000486
3	0.03	0.04	0.001944
4	0.02	0.04	0.000864

6. Calculate the activation energy, E_a, of the investigated reaction, Mg (s) + 2HCl (aq) → $H_2(g)$ + $MgCl_2$ (aq), based on the compiled temperature data of different teams (Part II).

Information: The Arrhenius equation, $k = Ae^{-E_a/RT}$, gives a relationship between the temperature and the rate constant. A more useful form of the equation is: $\ln k = \ln A - \{y = mx + b\}$. The value of the gas constant, R, is 8.314 J/mol · K. Thus if you plot the compiled data (part II) in the form $\ln k$ vs. $1/T$, the value of E_a can be calculated from the slope of the resulting line.